VITAMIN C

Recent Aspects of its Physiological and Technological Importance

An industry–university co-operation Symposium organised under the auspices of the National College of Food Technology, University of Reading, on 2nd and 3rd April, 1974

THE SYMPOSIUM COMMITTEE

GORDON G. BIRCH, B.Sc., Ph.D., F.R.I.C., M.R.S.H., F.C.S.
Reader at National College of Food Technology,
Reading University, Weybridge, Surrey.

J. A. FORD, B.A. (Oxon)
Senior Administrative Assistant at National College of Food Technology,
Reading University, Weybridge, Surrey.

K. J. PARKER, M.A., D.Phil. (Oxon)
General Manager, Tate & Lyle, Ltd., Group Research and Development,
Philip Lyle Memorial Research Laboratory,
Reading University, P.O. Box 68, Reading, Berks.

H. VIZARD ROBINSON, A.C.I.S.
Secretary at National College of Food Technology,
Reading University, Weybridge, Surrey.

VITAMIN C

Recent Aspects of its Physiological and Technological Importance

Edited by

G. G. BIRCH and K. J. PARKER

APPLIED SCIENCE PUBLISHERS LTD
LONDON

APPLIED SCIENCE PUBLISHERS LTD
RIPPLE ROAD, BARKING, ESSEX, ENGLAND

ISBN: 0 85334 606 2

WITH 62 TABLES AND 47 ILLUSTRATIONS

Printed in Great Britain by Galliard (Printers) Ltd Great Yarmouth

List of Contributors

A. E. BILLINGTON
Beecham Products, Beecham House, Great West Road, Brentford, Middlesex, England.

G. G. BIRCH
National College of Food Technology, University of Reading, Weybridge, Surrey, England.

B. M. BOINTON
National College of Food Technology, University of Reading, Weybridge, Surrey, England.

J. R. COOKE
Laboratory of the Government Chemist, Cornwall House, Stamford Street, London, England.

E. DEGKWITZ
Biochemical Institute, University of Giessen, Germany.

E. GINTER
Institute of Human Nutrition Research, Bratislava, Czechoslovakia.

D. M. GRESSWELL
Research and Development Department, Beecham Products, The Royal Forest Factory, Coleford, Gloucestershire, England.

J. D. HENSHALL
The Campden Food Preservation Research Association, Chipping Campden, Gloucestershire, England.

C. HINTON
The Campden Food Preservation Research Association, Chipping Camden, Gloucestershire, England.

D. HORNIG
F. Hoffmann-La Roche & Co. Ltd, Department of Vitamin and Nutritional Research, Basle, Switzerland.

R. E. HUGHES
University of Wales Institute of Science & Technology, King Edward VII Avenue, Cardiff, Wales.

H. KLÄUI
F. Hoffmann-La Roche & Co. Ltd, Department of Vitamin and Nutritional Research, Basle, Switzerland.

S. LEWIN
Department of Postgraduate Molecular Biology, North East London Polytechnic, Romford Road, London, England.

K. J. PARKER
Tate & Lyle, Ltd, Group Research & Development, Philip Lyle Memorial Research Laboratory, Reading University, P.O. Box 68, Reading, England.

M. D. RANKEN
Meat & Fish Products Section, The British Food Manufacturing Industries Research Association, Leatherhead, Surrey, England.

E. J. ROLFE
National College of Food Technology, University of Reading, Weybridge, Surrey, England.

J. D. SELMAN
National College of Food Technology, University of Reading, Weybridge, Surrey, England.

I. M. SHARMAN
Dunn Nutritional Laboratory, University of Cambridge, Milton Road, Cambridge, England.

Hj. STAUDINGER
Biochemical Institute, University of Giessen, Germany.

B. H. THEWLIS
Flour Milling & Baking Research Association, Chorleywood, Rickmansworth, Herts, England.

C. L. WALTERS
The British Food Manufacturing Industries Research Association, Leatherhead, Surrey, England.

C. W. M. WILSON
Department of Pharmacology, University of Dublin, Trinity College, Dublin, Ireland.

Contents

Opening Remarks

E. J. ROLFE

*National College of Food Technology,
University of Reading, Weybridge, Surrey, England*

A substantial amount of work, much of it empirical in nature, has implicated vitamin C in a wide range of apparently unrelated aspects of health and disease. Except for some understanding of the mode of action in the prevention of scurvy, our knowledge of its biochemistry and metabolism at cellular level is very small. The papers presented at this Symposium provide a valuable résumé of present thinking and knowledge of those actively working in the field. Vitamin C is of importance in many aspects of food processing, both in terms of retention of nutrition and its special chemical activity. This brief paper serves to introduce the more detailed treatment of specific topics which follow.

During the middle ages scurvy caused a heavy death toll amongst sailors. It was not until 1753 that James Lind, a Scottish naval surgeon, published the results of his therapeutic trials which demonstrated the remarkable curative effects of citrus fruits. He also pointed out that 'as oranges and lemons are liable to spoil . . . the next thing to be proposed is the method of preserving their virtues entire for years in a convenient and small bulk'. He recommended the evaporation of the fruit juice in a clean open earthen vessel, well glazed, for several hours on a water bath until the juice is the consistency of a syrup. 'It is then to be corked up in a bottle for use . . . so that thus the acid, and virtues of twelve dozen lemons or oranges may be put into a quart bottle, and preserved for several years'.[1] By following Lind's procedure a reasonable retention of vitamin C could be expected. However, nearly fifty years were to elapse before any appreciable advantage was taken of his work. This is perhaps not surprising, though one is surprised to learn that 'in

1912 when Scott set out for the South Pole he firmly believed that fresh meat was a good antiscorbutic, so loaded up with plenty of meat and what he thought were good strengthening foods. When Scott's men succumbed there was only a distance of 11 miles between them and safety. Had his rations contained sufficient vitamin C history might have been written differently'.[2] Perhaps Scott was relating his diet to that of Eskimos who appear to be capable of maintaining a healthy existence under conditions of extreme cold by eating meat and fish.

In spite of vitamin C being a relatively simple molecule we have little information concerning its role in tissue metabolism and are without an explanation for its apparently haphazard distribution in both plant and animal kingdoms. Some fruits and vegetables contain large amounts, *e.g.* blackcurrants, watercress, green peppers, whilst others contain relatively little, *e.g.* pears, plums, beetroot. It is a mystery why in some animal organs vitamin C is present in high concentrations, *e.g.* the adrenals, eye lens, vitreous and aqueous humour, white blood cells—and not in others, *e.g.* red cells, muscle, brain and pancreas. It is suggested that white blood cells provide storage and transport.

Most animal species are able to synthesise their vitamin C requirements, the exceptions are the primates, guinea-pigs, fruit eating bats and, so it is reported, the red-vented bulbul bird of India. The biosynthetic pathway is D-glucose → D-glucono γ-lactone → L-gulono γ-lactone → L-ascorbic acid. In species requiring ascorbic acid in the diet the liver lacks the enzyme necessary for the final step[3] L-gulonolactone oxidase. Primates—together with the Dalmatian dog—possess another biochemical lesion, *i.e.* the absence of the enzyme uricase. The final metabolite of ingested purines is uric acid. Proctor[4] points out that a number of the physiological functions of ascorbate are generally considered to be related to the unique electron donor properties of this compound. Uric acid (and purines) is also a strong electron donor. It is therefore possible that in the primates uric acid has taken over some of the functions of ascorbate, *i.e.* there might be a selective advantage in the loss of uricase in primate strains which had previously lost the ability to synthesise ascorbic acid. Recent work has indicated that vitamin C is rapidly lost from the tissues of guinea-pigs deprived of a dietary source of the vitamin, whereas in man the degree of depletion of vitamin C must be great in order that a wound fails to heal, *e.g.* after six months on a grossly deficient diet.[5] Is it possible that uric acid as proposed by Proctor is able to mitigate the effects of the deficiency specifically for man?

We have had the excitement of the discovery of the importance of protein, vitamins, etc., in the diet, and now we face the difficult task to determine the amounts needed to maintain full health, and to ascertain how well these amounts are satisfied from the food eaten. Surveys which rely on evaluation of adequacy of diets by relating food purchased with tables of vitamin C content of foodstuffs must be treated with caution. The amount of vitamin ingested is very dependent on the cooking method, and the extent to which the food is subsequently kept hot before consumption as may occur in restaurants and hopefully not in institutional feeding. The intake will also be derived only from the actual amount of food eaten and not the amount of food on the plate. Appropriate adjustments must always be made for food wastage and loss of nutrients during cooking and serving.

A number of concepts may be applied in assessing desirable intakes, *e.g.* the amount to satisfy the minimal needs of the individual, or amounts adequate to maintain the good nutritional state of particular groups of the population, or a recommended dietary allowance which will be adequate to cover individual variations in a substantial majority of the population. In order to maintain good health or nutritional state we may adopt either of two approaches (1) the person should be fully saturated with vitamin C, or (2) as long as the person appears to be in good health and any small wound heals normally, then he or she is probably getting all the vitamin C required.

The Department of Health and Social Security issued a report in 1969 entitled 'Recommended Intakes of Nutrients for the U.K.' It represents the judgement of an Expert Panel appointed by the Committee on Medical Aspects of Food Policy in reviewing and revising the allowances recommended by the Committee on Nutrition of the British Medical Association 1950. The Panel's recommendations take account of age, sex and occupation. The Panel preferred 'recommended intake' to dietary allowance, the former being defined as 'the amounts sufficient or more than sufficient for the nutritional needs of practically all healthy persons in a population'. It is therefore greater than the requirements for most individuals and thus provides a built-in safeguard. The recommended intake of ascorbic acid for adults is 30 mg/day. More information is given in Table 1, where recommended intake is related to the recommendations for

TABLE 1
Recommended daily intakes of vitamin C for the UK

	Body wt. (kg)	Ascorbic acid (mg)	Recommended intake Ascorbic acid (mg/1000 kcal)
Boys and Girls			
0 up to 1 year	7·3	15	19
1 up to 2 years	11·4	20	17
2 up to 3 years	13·5	20	14
3 up to 5 years	16·5	20	13
5 up to 7 years	20·5	20	11
7 up to 9 years	25·1	20	10
Boys			
9 up to 12 years	31·9	25	10
12 up to 15 years	45·5	25	9
15 up to 18 years	61·0	30	10
Girls			
9 up to 12 years	33·0	25	11
12 up to 15 years	48·6	25	11
15 up to 18 years	54·1	30	13
Men			
18 up to 35 years: sedentary	65	30	11
very active	65	30	8
35 up to 65 years: sedentary	65	30	12
very active	65	30	8
Women			
18 up to 55 years: most occupations	55	30	14
55 up to 75	53	30	15
Pregnancy, 2nd and 3rd trimester		60	25
Lactation		60	22

energy and which may be useful for certain purposes of planning as in practice it is the intake of energy that determines the intake of nutrients.[6] The UK recommendations are in line with the results of the Sheffield experiment on human volunteers to induce vitamin C deficiency. By using as the criterion the presence of scurvy it was found that the minimum protective dose was in the region of, or perhaps a little below, 10 mg/day.[7] The American recommendations are very much higher than those of the UK, being 60 mg for men and 55 mg for women,[8] in the belief that saturation is desirable because animals which synthesise the vitamin maintain their tissues at saturation level.

Pauling has claimed that substantial daily doses of vitamin C have proven to be beneficial in the prevention of the common cold.[9] Supporting evidence for this claim is from a recent experiment using 90 subjects one half of whom received 1 g/day ascorbic acid over a period of 15 weeks. Incidence of colds in the treated group was reduced by 49%, the reduction being significant at the 0·2% level.[10] Another experiment in Canada failed to show a reduced incidence of illness in subjects taking 1 g/day of vitamin C, but the group had 30% fewer days off work.[11]

To provide scientific support for a large daily intake of vitamin C the Stone–Pauling hypothesis assumes that the rate of biosynthesis of essential compounds is linked to physiological requirements. The rat synthesises vitamin C at the rate of about 50 mg/kg body weight daily, and according to the Stone–Pauling theory this would indicate a daily intake of 3–4 g for a 70 kg man. However, tissue concentrations in animals producing their own vitamin C are not significantly different from those of a man with a daily intake of 50–100 mg.[12]

Nutritional adequacy must take account of significant changes in human behaviour. The observation from animal experiments that oestrogens increase the rate of ascorbic acid breakdown and lower tissue levels[13] prompted an investigation of levels of ascorbic acid in blood cells from healthy women—some pregnant and some who were taking oral contraceptives.[14] Results showed that ascorbic acid in both leucocytes and platelets is significantly lower in women taking contraceptives than in the controls and pregnant women. It seems likely that oral contraceptive steroids increase the breakdown of ascorbic acid perhaps by stimulant action on the liver release of cerulo plasmin, a copper containing protein with ascorbate oxidase activity.[15] Many women taking oral contraceptives may be in an induced hypovitamin C condition and require supplementation.

Tissue levels of the vitamin are influenced by its origin, *e.g.* cabbage as a natural source of vitamin C produces higher levels than an equivalent quantity of pure vitamin administered in distilled water.[16] One explanation is the limited ability of the stomach to absorb the vitamin and whereas a large single dose would result in substantial losses due to non-absorption, taken as cabbage the vitamin is allowed to 'trickle' in over a prolonged period and leads to greater absorption and higher tissue levels.[17] Another possible contributory factor is that cabbage and other plant sources of ascorbic acid contain compounds that either increase the deposition of ascorbic acid in animal tissues or exert a sparing action for the vitamin. Bioflavonoids have been implicated in this respect. Rutin added to the diet of guinea pigs receiving sub-optimal amounts of ascorbic acid led to an increase in the levels of vitamin in the adrenals though not in the liver.[18] Similar results have been obtained using other flavonoids such as catechin.[19] It has also been shown that rutin is capable of prolonging the life span of guinea pigs on a scorbutogenic diet.[20] Following these ideas Hughes and Jones compared the tissue deposition of ascorbic acid and the growth response obtained using two natural sources of vitamin C (blackcurrant juice concentrate—rich in both vitamin C and bioflavanoids and acerola cherry juice powder—rich only in vitamin C) with those resulting from the administration of an equal amount of ascorbic acid.[21] The fruit juices were dosed orally under the same conditions as the pure ascorbic acid thus eliminating any differences in the rate of ingestion of the vitamin. The ascorbic acid concentrations mg/100 g tissue for the treatments, ascorbic acid solution, acerola cherry juice and blackcurrant juice were (1) in adrenals 17·0, 22·4 and 24·2 and (2) in spleen 5·74, 6·87 and 8·42 respectively. Surprisingly also body weight increases followed a similar pattern. The data suggest that flavonoids could influence absorption and/or the stability of the ascorbic acid in the tissue. Of greater significance is the effect on growth rates which is not attributable to differences in tissue ascorbic acid concentrations.

Little is known of the relationship between flavonoids and vitamin C in vegetable tissues. Autoxidation of lipids is catalysed by metals, particularly copper, and antioxidants may exert a protective effect by either chelating with the contaminating metal or by accepting free radicals and removing them from the system. Some flavonoids are effective in the protection of lipids; and it has been shown that some of the flavonoids isolated from blackcurrants are effective in protecting ascorbic acid from oxidation in aqueous solution, the most effective being quercetin which does not protect by chelating with copper.[22]

The ortho-hydroxylation compounds were most effective for protecting ascorbic acid and a similar finding has been reported on flavonoids as fat antioxidants. The stability of ascorbic acid in lemon juice can be increased by the addition of quercetin which also confers stability to flavour. However, we are far from understanding the role of the flavonoids in plant tissues.

The food industry is well aware of the importance of retaining the trace nutrients in processed foods, including vitamin C. During harvesting and handling some losses of ascorbic acid may occur, particularly due to enzymic action where tissue is damaged by cutting, bruising, etc. Subsequent processing and storage conditions can also give rise to further losses and in many cases, *e.g.* frozen peas it is found that retention of ascorbic acid is a reliable objective test for quality.

Because of its peculiar properties, ascorbic acid is finding application in diverse parts of the food industry and some interesting applications of this kind are to be discussed at this Symposium.

REFERENCES

1. Lind, J. (1753). *Treatise on Scurvy Part II*, Chapter IV, A. Kincaid and A. Donaldson, Edinburgh.
2. Norris, P. E. (1960). *About Vitamins*. Thorsons Publishers Ltd., London.
3. Horowitz, H. H., Doevschuk, A. P. and King, G. C. (1952). *J. Biol. Chem.*, **199,** p. 193.
4. Proctor, P. (1970). *Nature, Lond.* **228,** p. 868.
5. Crandon, J. H., Lund, C. C. and Dill, D. B. (1940). *New England J. Med.*, **233,** p. 353.
6. Greaves, J. P. (1969). *Nutrition.*, **23,** p. 124.
7. Krebs, H. A. (1953). *Proc. Nutr. Soc.*, **12,** p. 237.
8. National Research Council (1968). *Recommended Dietary Allowances*, a report of the Food and Nutrition Board, 7th rev. ed., National Academy of Sciences (National Research Council Publication No. 1694, Washington D.C.).
9. Pauling, L. (1970). *Vitamin C and the Common Cold*, W. H. Freeman, San Francisco.
10. Charleston, S. S. and Clegg, K. M. (1972). *Lancet*, **1,** p. 1401.
11. Anderson, T. W., Reid, D. B. W. and Beaton, G. H. (1972). *Can. Medical Assoc. J.*, **107,** p. 503.
12. Hughes, R. E. (1973). *British Nutrition Foundation Bulletin, No.* 9, p. 24.
13. Clemetson, C. A. B. (1968). *Lancet* (ii), p. 1037.
14. Briggs, M. and Briggs, Maxine (1972). *Nature, Lond.*, **238,** p. 277.
15. Osaki, S., McDermott, J. A. and Frieden, E. (1964). *J. Biol. Chem.*, **239,** p. 3570.
16. Kuether, C. A., Telford, I. R. and Roe, J. H. (1944). *J. Nutr.*, **28,** p. 347.
17. Penney, J. R. and Zilva, S. S. (1946). *Biochem. J.*, **40,** p. 695.
18. Douglass, C. D. and Kamp, G. H. (1959). *J. Nutr.*, **67,** p. 531.

19. Coterau, H., Gabe, M., Gero, E. and Parrot, J. L. (1948). *Nature, Lond.*, **161,** p. 557.
20. Abrose, A. M. and De Eds, F. (1949). *J. Nutr.*, **38,** p. 305.
21. Hughes, R. E. and Jones, P. R. (1971). *J. Sci. Fd Agric.*, **22,** p. 551.
22. Clegg, K. M. and Morton, A. D. (1968). *J. Fd Technol.*, **3,** p. 277.

1

Vitamin C: Historical Aspects

I. M. SHARMAN

Dunn Nutritional Laboratory, University of Cambridge and Medical Research Council, Milton Road, Cambridge, England

ABSTRACT

Much of the early history of vitamin C is concerned with the aetiology, treatment, and prevention of the dreaded disease scurvy. There was a time when this disease was responsible for more losses amongst naval personnel than from all other causes including hostilities and shipwreck. James Lind, the Scottish naval physician (1716–94) *played a prominent part in recognising the antiscorbutic value of citrus fruits. He carried out the first clinical trial that showed conclusively the therapeutic value of lemon juice in curing the disease. The results of this trial were included in his famous* A Treatise of the Scurvy *published in* 1753 *in which he recommended the issue of a daily ration of the juice. Nearly half a century, however, was to elapse before his proposals were adopted in* 1795 *at the instigation of Sir Gilbert Blane. A dramatic fall in the incidence of the disease then ensued, and by* 1800 *the conquest of scurvy was practically complete. However, during the nineteenth century there appear to have been further outbreaks of the disease because Lind's recommendations did not continue to be implemented.*

Modern work on scurvy began in 1907 *when Holst and Frölich produced the disease in guinea-pigs by giving them a restricted diet. In* 1912 *Hopkins showed that animals could not thrive on purified isolated food constituents and he introduced the concept of 'accessory food factors'. In the same year Casimir Funk postulated the existence of 'vitamines' and subsequently suggested the absence of different vitamins was responsible for the onset of different 'deficiency' diseases. By* 1924 *Zilva and his colleagues effected a* 300*-fold concentration of the active compound from lemon juice. Szent-Györgyi in* 1928, *during a course of studies on cellular oxidation, isolated what he called 'hexuronic acid', from cabbage and adrenals. Four years later Waugh and King isolated the vitamin from lemons and showed it to be identical with the 'hexuronic acid' of Szent-Györgyi; they also showed that the anti-scorbutic activity of 'hexuronic acid' was similar to that of vitamin C obtained from orange juice. The vitamin was finally identified and given the name ascorbic acid in* 1933 *in which year it was also synthesised by various authors simultaneously.*

EARLY HISTORY OF SCURVY

The early history of vitamin C is intimately connected with the aetiology, treatment, and prevention of the terrible and dreaded disease scurvy. This disease is an ancient one and differs in this respect from the other vitamin deficiency diseases such as beriberi and pellagra which have been discovered more recently. For centuries scurvy was the cause of more deaths amongst naval crews than resulted from all other causes including battle and shipwreck.

Scurvy was first described with certainty at the time of the Crusades in the thirteenth century; for hundreds of years it was a common occurrence not only amongst sailors and explorers but also among normal populations in Northern Europe. Roddis[1] has drawn attention to the profound effects that followed the change from oars to sail as a method for the propulsion of ships. It was the full development of the mast and sail that made long oceanic voyages possible. The oar-propelled ship was suitable only for short excursions along the coast or in inland waters and was unable to cope with the heavy swells and rough weather encountered on the open sea. With the sailing ship came increase in size and a more complete compartmenting below decks. Coasting vessels were away from land for only a few days and consequently were able to obtain adequate fresh provisions, with the result that diseases such as scurvy were of minor importance until the long open sea voyage became common and the necessity arose for the use of salted and preserved foods. Furthermore the larger ships with the increased number of decks and compartments resulted in inadequate ventilation and an increased incidence of respiratory and contagious diseases.

The symptoms of scurvy were very definite and characteristic; the onset generally being accompanied by weakness and lassitude. This was followed by swelling of the legs and arms, softening of the gums, and haemorrhages. Nose bleed and bleeding from the gums were usually prominent features and then there were the haemorrhages under the skin that produced the characteristic ecchymoses. As the case advanced the teeth became loose and not infrequently fell out. Another symptom was foulness of breath and general strength failed so much that the patient was unable to stand. If the disease was not relieved death would result from exhaustion or from acute infections such as pneumonia. Such infections usually produced a rapid fatal ending due to the patient's debilitated state. A typical feature was the frequent sudden death from acute heart failure which was seen even in young men.

Some idea of the havoc that the disease caused may be gleaned from allusions to the subject that appear in contemporary records. Thus Harris[2] recalls that when Vasco da Gama sailed round the Cape of Good Hope in 1498 one hundred men out of a crew of 160 are said to have perished from the disease. A Spanish galleon was also reported adrift at sea with its entire crew dead from scurvy. In 1593 Admiral Sir Richard Hawkins referred to 10 000 seamen having died from scurvy within his personal experience. It is clear from accounts such as these that the disease was a scourge which was justifiably dreaded and abhorred.

During Jacques Cartier's second voyage to Newfoundland in 1535 scurvy broke out amongst the crew. Many of the sailors died but one of them learned from the Red Indians that a decoction made from the needles of spruce trees was an effective cure. Such a concoction was prepared and when put to the test produced almost miraculous results, the disease disappearing almost immediately. Records of other cures are also known and in 1593 Sir Richard Hawkins effected a cure by giving lemon juice to his men. Another striking example of early knowledge regarding the prevention of scurvy is to be found in the account of the first voyage sent out to the East Indies in 1600 by the newly established East India Company. A squadron of four ships was sent under the command of Captain James Lancaster who was in the *Dragon*. When the ships arrived at the Cape of Good Hope, all the crews except those aboard the *Dragon* were so depleted by scurvy that they were scarcely able to bring their ships to anchor. However, the flagship had no such experience due to the use of lemon juice which Captain Lancaster had brought to sea in bottles and of which he issued three teaspoonsful every morning to those who showed the slightest sign of scurvy. When the ships were finally anchored Lancaster is reported to have used lemon juice as a restorative to the crews of the other ships. Of a total of 424 men 105 were lost on this expedition but all those who died were aboard the lesser ships of the squadron.

It is likely that John Woodall was well aware of Lancaster's experiences with lemon juice. In 1617 when he published his book *The Surgeon's Mate* he strongly advised the use of lemon juice both for the prevention and treatment of scurvy. Here are his own words: 'The use of the juice of Lemmons is a precious medicine and well tried, being sound and good, let it have the chief place, for it will deserve it, the use whereof is; It is to be taken each morning, two or three spoonfuls, and fast after it two hours, and if you add one spoonful of Aquavitae (brandy) thereto to a cold stomach, it is the better. Also if you take a little thereof at night it is good to mix therewith some sugar or to take of the syrup thereof is not amiss. Further note it is good to be put into each purge you give in that disease. Some Chirurgeons also give of this juice daily to the men in health as a preservative, which course is good if they have store, otherwise it were best to keep it for need. I dare not write how good a sauce it is at meat, lest the chief (chef) in the ships waste it in the great Cabins to save vinegar. In want whereof use the juice of Limes, Oranges, or Citrons, or the Pulp of Tamarinds'. Here then we have an early and indisputable statement that the juices of the citrus fruits can prevent scurvy.

In 1696 Dr William Cockburn, in his book *Sea Diseases, or Treatise of their Nature, Cause and Cure*, clearly pointed out the value of fresh fruits and vegetables in the treatment of scurvy. There were, however, many other suggestions for treating the disease; these included elixir of vitriol, vinegar, salt water, cinnamon, and whey. Nevertheless medical men of the time were agreed on the effects of exposure, fatigue, and depression in precipitating the attacks and on the need for rest and warmth in their treatment. A final and irrevocable proof of the efficiency of lemon juice, however, was not forthcoming until the middle of the next century.

THE CONQUEST OF SCURVY

James Lind, the Scottish naval surgeon of the eighteenth century (1716–94) has been called both the 'father of naval hygiene' and the 'Hippocrates of naval medicine'. He was one of the foremost pioneers in preventive and tropical medicine, and played a distinguished part in finding the cause of scurvy and in suggesting a means for its eradication. Lind was born in Edinburgh in 1716 and at the age of fifteen was apprenticed to Dr Langlands, a prominent physician in that city. In 1739 he entered the Royal Navy as a surgeon's mate and a few years later was promoted to the rank of surgeon. It was in this capacity when on board the 74-gun ship *Salisbury* in 1747 that he carried out what may be regarded as the first controlled clinical trial which showed conclusively that fresh citrus fruits will cure scurvy. Here is his own account of what happened:

'On the 20th May 1747, I selected twelve patients in the scurvy, on board the *Salisbury* at sea. Their cases were as similar as I could have them. They all in general had putrid gums, the spots and lassitude, with weakness of their knees. They lay together in one place, being a proper apartment for the sick in the fore-hold; and had one diet common to all, *viz.* water-gruel sweetened with sugar in the morning; fresh mutton-broth oftentimes for dinner; at other times light puddings, boiled biscuit with sugar, etc., and for supper, barley and raisins, rice and currants, sago and wine, or the like. Two of these were ordered each a quart of cyder a-day. Two others took twenty-five drops of elixir vitriol three times a-day, upon an empty stomach; using a gargle strongly acidulated with it for their mouths. Two others took two spoonfuls of vinegar three times a day, upon an empty stomach; having their gruels and their other food sharpened with vinegar, as also the gargle for their mouth. Two of the worst patients, with the tendons in the ham quite rigid (a symptom none of the rest had) were put under a course of sea water. Of this they drank half a pint every day, and sometimes more or less, as it operated, by way of gentle physic. Two others had each two oranges and one lemon given them every day. These they eat with greediness, at different times upon an empty stomach. They continued but six days under this course, having consumed the quantities that could be spared. The two remaining patients took the bigness of a nutmeg three times a-day, of an electary recommended by an hospital-surgeon, made of garlic, mustard-seed, horse-radish balsam of *Peru*, and gum myrrh; using for common drink barley-water boiled with tamarinds; by which, with the addition of *cream of tartar*, they were gently purged three or four times during the course.'

The results of these different attempted remedies, were reported by Lind as follows:

'The consequence was, that the most sudden and visible good effects were perceived from the use of oranges and lemons; one of those who had taken them, being at the end of six days fit for duty. The spots were not indeed at that time quite off his body, nor his gums sound; but without any other medicine than a gargle for his mouth he became quite healthy before we came into Plymouth which was on the 16th of June. The other was the best recovered of any in his condition; and being now pretty well, was appointed nurse to the rest of the sick.'

Lind reported his results in his famous book *A Treatise of the Scurvy*, which was published in 1753[3] and from which the above extracts have been taken. It is clear from Lind's observations that had the issue of lemon juice to sailors on long distance voyages been brought in at once the disease might well have been conquered immediately. However, despite the convincing proof of the efficacy of lemon juice for curing scurvy it was to be a long time before Lind's proposals were universally adopted and thousands of seamen were to die meanwhile.

Some forty years were in fact to elapse after the publication of Lind's treatise before the issue of lemon juice in the navy was made compulsory. In 1795 Sir Gilbert Blane was appointed one of the two medical commissioners of the Board of the Sick and Hurt. In the same year, through his recommendations, Lord Spencer who was then First Lord of the Admiralty put into effect the forced issue of lemon juice to all ships. A marked reduction in the incidence of scurvy ensued and by 1800 the conquest of the disease was virtually complete.

However, later on in the nineteenth century the disease again broke out and this was evidently due to the fact that Lind's recommendations did not continue to be fully implemented. Lessons which had been learned were forgotten and mistakes were made continually. Because the small green lime was grown extensively in the British West Indies, the lemon juice ration was often referred to as lime juice, and it was this that led to the slang term 'lime-juicer', or more frequently 'limey', being applied to British ships and sailors. In the Royal Navy and subsequently in the merchant service the juices of the various citrus fruits were all called lime juice, but almost invariably it was the juice of the lemon (*Citrus medica*, var. *limonum*) that was issued. The lemons that were used came from the Mediterranean region, particularly Sicily, Malta, Lisbon and Madeira. In 1845 the Governor of Bermuda, with an eye, no doubt, to creating a market for the local product suggested to the Admiralty that the West

Indian lime (*Citrus medica*, var. *acidum*) might be used as a substitute for the lemon. Gradually the Admiralty contracts were altered and lime juice was supplied instead of lemon juice. However, the substitution led to tragic happenings of which the cause was only discovered many years later, when scurvy and antiscorbutics were being investigated experimentally. It is now known that the vitamin C content of the West India lime is only about one-half that of the orange or lemon. Typical recent values of vitamin C in the common citrus fruits are shown in Table 1. As there was an appreciable loss

TABLE 1

Vitamin C in citrus fruits

	Vitamin C (mg/100 g)
Lemons, or lemon juice	50
Oranges	50
Grapefruit	40
Tangerines	30
Limes	25

in the antiscorbutic effect of both varieties of juice it can be understood that the amount provided by the lime juice could have been less than the body's needs and it was for this reason that further outbreaks did occur.[4]

Nowadays, because of our understanding of the need for the antiscorbutic factor and of the vitamin C content of different foods it is possible to provide a diet containing sufficient of the vitamin and the disease is therefore practically unknown. Before proceeding to an account of the unravelling of the nature of the antiscorbutic factor we should look further at the incidence of scurvy in voyages and expeditions.

SCURVY IN VOYAGES AND EXPEDITIONS

Although, as has already been said, many years were to pass after the publication of Lind's treatise before the issue of lemon juice was universally made there were some notable examples of the use of the juice and of fresh foods in preventing the onset of scurvy. Thus

Captain James Cook on his first voyage to Australia in 1770 successfully made use of the fact that fresh fruit and vegetables would indeed prevent scurvy. During his second voyage round the world, which took place from 1772 to 1775, he lost only one man from sickness and that not from scurvy. He took every opportunity when touching land to obtain supplies of fresh fruit and vegetables. He believed in the virtues of celery and scurvy-grass and always had on board an ample provision of *Sauerkraut*, which had previously been found in the Dutch navy to be beneficial (*see* Table 2). He also took

TABLE 2

Vitamin C in sauerkraut and cabbage

	Vitamin C (mg/100 g)	
Sauerkraut	20	
	Raw	Cooked
Spring cabbage	80	25
Winter cabbage	55	15
White cabbage	40	—

with him a good supply of malt from which was brewed a 'sweet-wort', served liberally to any man showing signs of scurvy; a 'rob' of oranges and lemons was also kept available.[5]

In February 1776 Cook was elected a fellow of the Royal Society and the following month read to the Society a paper on his methods of preventing and curing scurvy which incidentally constitutes by no means the least of his services to Britain. It is interesting that the Royal Society subsequently awarded him the Copley gold medal, the highest honour of the Society, for his practical demonstration of the means of preventing scurvy rather than for his remarkable feats of navigation. Kodicek and Young[6] have suggested that the significance and uniqueness of the contribution that Cook made was 'not so much that he used greens and fruit as antiscorbutics, since these had been advocated before, but by his passionate determination to enforce by example and by authority on his sometimes non-cooperative crew and officers, a dietary regime which he believed was right'.

The incidence of scurvy in British polar expeditions between 1875 and 1917 has been reviewed by Kendall.[7] Those expeditions whose diet provided little of what we now know as vitamin C during the winter experienced scurvy early on the spring sledge journeys. However, during the period reviewed the food eaten increased in variety and this resulted in an increase in the vitamin C content of the diet as a whole. It also became recognised that game obtained during a journey significantly added to the vitamin intake. As the vitamin C rose fewer cases occurred and those which did occur could be explained by the special individual circumstances of the men concerned. Although modern expeditions are never likely to encounter scurvy again to the degree formerly encountered the experience of the past would still be a valuable asset should such a calamity befall a modern party.

TOWARDS A RECOGNITION OF THE ANTISCORBUTIC FACTOR

Although Lind had shown conclusively that lemon juice can cure scurvy there was no suggestion in his writings that the disease was caused by the absence of a single principle or nutrient. During the nineteenth century some progress was made towards the recognition that this is indeed the case. Thus George Budd (1808–82) described in considerable detail three of the deficiency diseases, viz. scurvy and what are now known as xerophthalmia and rickets. When physician to the Dreadnought Seamen's Hospital Ship, Budd had extensive first-hand experience of scurvy. In 1842 he published a series of five articles based on his lectures at King's College and called 'Disorders resulting from defective nutriment'.[8] In these articles Budd put forward his description of scurvy as a nutritional disorder, *e.g.* he said 'Men had not yet perceived that the disease had its real origin, not in the cold of our rigorous climate but in the abstinence from fresh vegetables and fruits'. The suggestion is made quite clearly that accessory food factors were necessary parts of normal diets and furthermore that distinct diseases would follow their absence. In this respect his ideas preceded the emergence of the currently accepted theory of vitamins and their corresponding deficiency diseases. Commenting on Budd's foresight into the supposed aetiology of these diseases Hughes[9] has said 'It is a measure of Budd's perceptive powers that he not only recognised them as diseases of nutritional deficiencies but that he also introduced (albeit abortively) into nutritional thought the concept that a specific disease could result from the absence of a single dietary component'.

Budd also forecast that chemists would soon isolate the antiscorbutic substance, the 'essential element' as he called it, from natural products. In actual fact it was not until 1907 that any significant advance in this direction was made. In that year Holst and Frölich[10,11] produced scurvy in guinea-pigs by feeding them a restricted diet. This was the first time that animals had been used in experimental investigations of scurvy. These trials were undertaken at the request of the Norwegian Government who were concerned at the time with the repeated occurrence of cases of the so-called 'ship beriberi' in both the Norwegian navy and the mercantile marine. Holst and Frölich were able to show that the removal of the customary fresh green foods from the diet of guinea-pigs, and of their subsistence on a cereal ration of oats, rye and rice, caused a fatal disease which closely resembled human scurvy. The signs observed included loss of weight, loosening of the teeth, haemorrhages in various parts of the body and extensive bone lesions; these signs were therefore similar to those seen in the human. Because of the many similarities seen in the guinea-pigs the disease was considered to be analogous to human scurvy. During the course of their research Holst and Frölich were able to show that a daily ration of fresh cabbage, carrot, dandelion leaves, cranberries or fresh fruit juices had a protective and curative effect when added to the diet. The antiscorbutic potency was also reported to be reduced if the vegetables were dried. It was also shown that the prolonged cooking of vegetables reduced their antiscorbutic value but that the potency of fruit juices was unaffected; the latter was correctly attributed to their acid reaction. Finally these authors demonstrated that the antiscorbutic factor was insoluble in ethanol unless it was acidified, and that it passed easily through a dialysing membrane. Another interesting result obtained at about this period was the discovery that dry cereals and legumes, viz. oats, barley, peas, beans and lentils, which have very little antiscorbutic properties, develop them during germination.[12]

In 1912 came the classic demonstration by Hopkins[13] that animals could not thrive on purified isolated food constituents but that they did when given small supplements of milk. From the results of his experiments Hopkins introduced the concept of 'accessory food factors'. In the same year Casimir Funk[14] postulated the existence of 'special substances which are of the nature of organic bases, which we will call vitamines'. When it was later realised that all vitamins were not vital amines, as had been supposed, the terminal 'e' was dropped at the instigation of Drummond.[15] Funk's 'scurvy vitamine' subsequently became 'water-soluble C' and finally, vitamin C, as we know it at present. It is, of course, by this name that our symposium to-day has been called.

EVENTS LEADING TO THE SYNTHESIS OF VITAMIN C

Early attempts to isolate the vitamin were seriously handicapped by the ease with which it can be destroyed by oxidation. Studies on the chemical nature of vitamin C were begun in 1918 by Harden and Zilva[16] at the Lister Institute. Another group under the direction of Nikolai Bezssonoff[17] in France took up the trail in 1921. Between 1924 and 1929 Zilva and his colleagues[18–20] were successful in obtaining a concentrate from lemon juice. They effected a 300-fold concentration of the active substance. Four years later in Hopkins' laboratory at Cambridge, Szent-Györgyi,[21] during investigations into cellular oxidation, isolated from cabbage and adrenals what he called 'hexuronic acid'. King[22] has described how, during a post-doctoral year from 1926 to 1927 he undertook the characterisation of vitamin C as a part of the graduate student training at the University of Pittsburgh. During the course of this work he and his collaborators followed three basic practices: (1) to assay all final fractions from each new series of laboratory tests, (2) to regulate very carefully the pH values of all solutions, and (3) to avoid, as far as possible, exposure to copper and air. No doubt the strict keeping to these practices aided King and his collaborators in the successful attainment of their goal.

While completing their thesis work Sipple[23] and Grettie[24] obtained during the years 1927–29 increased yields with higher activity than had ever been obtained before, and in less time. The active material was found invariably to behave as a rapidly diffusible anion, even during electrodialysis in a direct current.[25] In agreement with Zilva, King found that the molecular weight of the vitamin was about that of glucose. There was never any evidence of separation into more than one component. King reported these results at the spring meeting of the American Chemical Society in 1929. Dr E. C.

Kendall, who was present at the meeting, suggested to King that his material bore a remarkable resemblance to the 'hexuronic acid' that had just been reported by Szent-Györgyi. During the autumn of the same year, 1929, King again reported his findings, this time at a seminar at Cambridge University. Afterwards, Hopkins somewhat excitedly invited King into his office and asked him whether he would venture a guess as to the chemical identity of the vitamin. King replied that its properties and occurrence, so far as they were known, corresponded with the 'hexuronic acid' that had been isolated in Hopkins' laboratories. Hopkins then explained at some length that L. J. Harris had already called attention to the similarities between the acid and Zilva's preparations at a previous seminar and that he had sent a sample to Zilva in London. It seems, however, that Zilva never reported the results of his tests with the preparation and only replied by saying that it was not vitamin C.

Two further students of King, Smith[26] and Svirbely[27] simplified the concentration procedures further and reached an approaching constant activity by the summer of 1931. They were, however, unable to obtain a regular crystalline preparation. Finally, Waugh and King[28,29] succeeded in obtaining crystalline preparations fairly regularly and with constant assay activity of approximately 0·5 mg/day. Further reports of later assays soon followed from other laboratories and all these agreed that the minimum protective level was of the order of 0·5 mg/day. Other important contributions that closely followed the publication of Waugh and King's separation were those of Tillmans and Hirsch[30] who succeeded independently in crystallising the vitamin, and of Harris *et al.*[31] who confirmed that 'hexuronic acid' had antiscorbutic value. In 1933, Szent-Györgyi and his group[32] made another major contribution by discovering a method for preparing large quantities of the vitamin from paprikas. The availability of such relatively large quantities paved the way for the elucidation of the structure of the vitamin by a number of European organic chemists.

While investigations into the precise molecular structure were taking place Szent-Györgyi and Haworth[33] in 1933 proposed the name 'ascorbic acid' for the vitamin and this subsequently received universal acceptance.

It was not long before the structure of the vitamin was ultimately elucidated. Haworth and his colleagues[34,35] were the first to suggest the formula on the basis of identifying the derivative of L-threonic acid. They also correctly postulated the enolic lactone form of the molecule and they did so on the basis of reversible oxidation products and the methyl ethers at positions 2 and 3, and were thereby able to synthesise the vitamin. Other important contributions to our understanding of the structure were made by Karrer, Salomon, Schöpp, Morf and Zhender[36,37] and by Micheel and Kraft[38]. Reichstein and his colleagues[39,40] were especially successful in proposing methods for the synthesis of the vitamin which confirmed its structure on the basis of its preparation either from xylose or from sorbose. The way was then opened up for the commercial preparation of the vitamin from glucose. A large number of methods for the synthesis of the vitamin have been suggested and time does not allow a discussion of these, suffice it to say that the important steps for the industrial synthesis have been from glucose via sorbitol, sorbose, acetone–sorbose, 2-ketogulonic acid and eventual ring closure to the enolic lactone.

We have now traced the history of vitamin C from its early beginnings. We have seen what an anathema scurvy was in bygone days, we have considered the contributions made by Lind and others to the recognition of the antiscorbutic factor, and we have briefly outlined the stages in the ultimate synthesis of the compound. It will be left to other speakers at this Symposium to describe and inform us of more recent developments in our knowledge of this fascinating and intriguing compound.

REFERENCES

1. Roddis, L. H. (1950). *James Lind: Founder of Nautical Medicine,* Henry Schulman, New York, p. 31.
2. Harris, L. J. (1937). *Vitamins in Theory and Practice,* Cambridge University Press, Cambridge, p. 79.
3. Lind, J. (1753). *A Treatise of the Scurvy,* 1st Edition, Sands, Murray and Cochran, Edinburgh. For A. Kincaid and A. Donaldson.
4. Smith, A. H. (1918). *J. Royal Army Med. Corps,* **32,** pp. 93, 188.
5. Chick, H. (1953). *Proc. Nutrit. Soc.,* **12,** p. 210.
6. Kodicek, E. H. and Young, F. G. (1969). *Notes & Recs. Roy. Soc. Lond.,* **24,** p. 43.
7. Kendall, E. J. C. (1955). *Polar Record,* **7**(51), p. 467.
8. Budd, G. (1842). *Lond. Med. Gazette,* **2,** pp. 632, 712, 743, 825, 906.
9. Hughes, R. E. (1973). *Med. Hist.,* **17,** p. 127.
10. Holst, A. and Frölich, T. (1907). *J. Hyg. Camb.,* **7,** p. 634.

11. Holst, A. and Frölich, T. (1912). *Z. Hyg. Infektkr.*, **72,** p. 1.
12. Fürst, V. (1912). *Z. Hyg. Infektkr.*, **72,** p. 121.
13. Hopkins, F. G. (1912). *J. Physiol.*, **44,** p. 425.
14. Funk, C. (1912). *J. State Med.*, **20,** p. 341.
15. Drummond, J. C. (1920). *Biochem. J.*, **14,** p. 660.
16. Harden, A. and Zilva, S. S. (1918). *Biochem. J.*, **12,** p. 259.
17. Bezssonoff, N. (1921). *C.r. Acad. Sci., Paris,* **173,** p. 417.
18. Zilva, S. S. (1924). *Biochem. J.*, **18,** p. 182.
19. Zilva, S. S. (1924). *Biochem. J.*, **18,** p. 632.
20. Zilva, S. S. (1929). *Biochem. J.*, **23,** p. 1199.
21. Szent-Györgyi, A. (1928). *Biochem. J.*, **22,** p. 1387.
22. King, C. G. (1953). *Proc. Nutrit. Soc.*, **12,** p. 219.
23. Sipple, H. L. and King, C. G. (1930). *J. Amer. chem. Soc.*, **52,** p. 470.
24. Grettie, D. P. and King, C. G. (1929). *J. biol. Chem.*, **84,** p. 771.
25. McKinnis, R. B. and King, C. G. (1930). *J. biol. Chem.*, **87,** p. 615.
26. Smith, F. L. Jr. and King, C. G. (1931). *J. biol. Chem.*, **94,** p. 491.
27. Svirbely, J. L. and King, C. G. (1931). *J. biol. Chem.*, **94,** p. 483.
28. Waugh, W. A. and King, C. G. (1932). *J. biol. Chem.*, **97,** p. 325.
29. Waugh, W. A. and King, C. G. (1932). *Science,* **76,** p. 630.
30. Tillmans, J. and Hirsch, P. (1932). *Biochem. Z.*, **250,** p. 312.
31. Harris, L. J., Mills, J. I. and Smith, F. (1932). *Lancet,* **223,** p. 235.
32. Svirbely, J. L. and Szent-Györgyi, A. (1933). *Biochem. J.*, **27,** p. 279.
33. Szent-Györgyi, A. and Haworth, W. N. (1933). *Nature, Lond.*, **131,** p. 24.
34. Ault, R. G., Baird, D. K., Carrington, H. C., Haworth, W. N., Herbert, R., Hirst, E. L., Percival, E. G. Y., Smith, F. and Stacey, M. (1933). *J. chem. Soc.*, p. 1419.
35. Haworth, W. N., Hirst, E. L. and Smith, F. (1934). *J. chem. Soc.*, p. 1556.
36. Karrer, P., Salomon, H., Schöpp, K. and Morf, R. (1933). *Helv. chim. Acta,* **16,** p. 181.
37. Karrer, P., Schöpp, K. and Zhender, F. (1933). *Helv. chim. Acta,* **16,** p. 1161.
38. Micheel, F. and Kraft, L. (1933). *Hoppe-Seyl. Z.*, **222,** p. 235.
39. Reichstein, T., Grüssner, A. and Oppenauer, R. (1933). *Helv. chim. Acta,* **16,** p. 1019.
40. Reichstein, T. and Grüssner, A. (1934). *Helv. chim. Acta,* **17,** p. 311.

DISCUSSION

Rivers: Dr Sharman reported scurvy as first occurring in the thirteenth century, but there are said to be hieroglyphs in the Ebers papyrus referring to scurvy occurring in Ancient Egypt. Browthwell, in examination of Neolithic remains, points out that some of the bone changes are not incompatible with chronic vitamin C deficiency. Indeed, the evidence he presented about Red Indians pointing out to Jacques Cartier's men a cure for scurvy makes it very likely that it also existed in that hunting–gathering society.

I think possibly the reason might be—I should very much appreciate Dr Sharman's comments on this—the reason that the first detailed reports are in the Crusades and in sailing ships might be that at this point scurvy became of political and military importance, and the disease was therefore worth doing something about. In 1939, Le Gros Clark and Titmuss wrote a very good book called *Our Food Problem*, in which they pointed out that, had anyone cared to look, scurvy probably occurred every winter in and around many isolated villages in England throughout the Middle Ages, and that it wasn't purely a disease due to the invention of the sail. It was the invention of the sail that made it a disease of some importance in political, economic and military terms.

Sharman: Yes, I agree with most of what you say. I didn't mean to suggest that there had been no scurvy before the thirteenth century; what I did say, I think, was that this was the first correct account, the first account that could with certainty be attributed to scurvy. I am quite sure that, as you say, the disease did exist before.

Wilson: There are well-substantiated reports, are there not, Dr Sharman, of the use of scurvy-grass amongst the Vikings, and in fact the word 'scurvy' apparently is derived from the original Viking language, in which they described how they used it. It has been suggested that they went over to America via Greenland, because they picked up scurvy-grass on the way.

The second point I would like to ask Dr Sharman, as a point of information, is, can he tell us anything about the voyage of the *Mayflower*, and the birth of a child on that voyage?

Sharman: That is a very interesting point, but I am afraid I cannot give you any further information on it.

2

Technical Uses of Vitamin C

H. KLÄUI

F. Hoffmann-La Roche & Co. Ltd, Department of Vitamin and Nutritional Research, Basle, Switzerland

ABSTRACT

In view of the various other papers covering the more important special fields of application of vitamin C in food manufacture, only some more general aspects and some problems pertaining particularly to the technical uses of vitamin C are dealt with in this report.

After mentioning some figures showing the continuously increasing significance of vitamin C in food manufacture, the reasons why a relatively unstable compound like vitamin C should achieve such an important role are described.

A practical problem of general interest concerns the formation of nitrosamines in curing meat with nitrites, and the influence of vitamin C on residual nitrites and nitrosamines.

Some general precautions which should be observed when using ascorbic acid in foods, particularly beverages, are listed; these measures aim at maintaining a sufficient overage after bottling or canning, since this is essential for achieving a satisfactory effect.

A potentially very interesting application is the use of fatty acid esters of vitamin C in antioxidant mixtures for fats and oils. Some recent results are given which show the effectiveness of these antioxidants, together with the properties of the compounds themselves, particularly the poor solubility and low rate of dissolution of ascorbyl palmitate.

A few technical problems which may arise in connection with vitamin C enrichment of foods are mentioned.

Finally, some experimental work with ascorbic acid as an abscission agent is summarised. Inducing abscission of fruits by spraying trees with ascorbate solutions has been tried with olives and oranges. Factors such as humidity strongly influenced the results. The use of vitamin C as an abscission agent is potentially very interesting, since it could significantly improve the economics of mechanical harvesting.

INTRODUCTION

The total annual production of carotenoids is estimated to amount to about 100 million tons, and of this figure, the amount of carotenoids produced by synthesis is less than 1 ppm. To my knowledge, no estimate is available of Nature's production of vitamin C; although it is certainly much larger than the amount produced by synthesis, my guess is that the latter represents a significant part, and probably a quantity of the same order of magnitude as that naturally present in the food eaten by the whole world population.

The Nobel Prize winner, Professor Tadeus Reichstein, first achieved his synthesis of vitamin C in 1933, and the following year the industrial scale synthesis was begun by Roche in Basle following his fundamentals. As a result of the continuing development and growth, the price of the vitamin has been drastically reduced. In 1950, 1 kilogramme cost about 770 Swiss francs: 10 years later it had come down to around 129 Swiss francs, and today it is a little over 18 Swiss francs.

About half of the total amount of vitamin C produced synthetically is used today in the food industry—a comparatively small amount as a nutritional vitamin additive, and the largest portion as a technological aid. Obviously, it is not possible to draw a clear division between these two uses. Ascorbic acid is vitamin C: thus in foods it must have a dual role. For the purposes of this lecture the division is made by considering the reason for its use. If it is added to enrich the product's vitamin C content, it is used as a vitamin: if the addition is made to achieve some improvement in, for example, quality, shelf-life or processing, without regard for its vitamin activity in the end product, this is a technological use.

The proportion of ascorbic acid used for technological purposes in foods has been increasing steadily and here, perhaps, I should say that I have interpreted the title of my paper as 'Technical Uses of Vitamin C in Food Manufacture' and therefore I am not going to speak about vitamin C in photography, plastics manufacture, hair waving preparations, stain removers or water treatment (for the elimination of excess chlorine). While these applications are good examples of the versatility of ascorbic acid they are not food uses.

FOOD APPLICATIONS

The first food applications were in beer (24 years ago), in meat and preserves (20 years ago), and, towards the end of the 1950s, in flour to improve baking qualities and in soft drinks and wine to act as a stabiliser.[1] Due to the relatively high cost and limited availability of vitamin C at those times, its applications in the food industry were insignificant and the main outlet was in pharmaceutical preparations. Increased production in the 1960s, coupled with lower costs, allowed a broader technical use: 10 years ago it was estimated to be one quarter, and today it is believed to reach almost one half of the total world production. This development has led to the possibility of reducing the number or quantity of non-physiological additives such as sulphurous acid and nitrites.

The main fields of application[2] are now:

soft drinks, especially beverages based on citrus fruits, where vitamin C functions as an antioxidant for the flavours as well as a nutrient;
meat and *meat-containing products*, in meat curing and pickling;
flour, for improving the baking quality; and
beer, as a stabiliser.

I mention these facts to show the significance of vitamin C in food manufacture, particularly its technical uses, and the various developments which have led to a continuously increasing consumption.

The question arises why a vitamin, and especially a relatively unstable compound like vitamin C, should achieve such an important role in food manufacture. Several factors have to be considered, and their sum explains the unique position of this product.

The chemical, and especially the reducing, properties of vitamin C have made this compound useful as an antioxidant and stabiliser, and as a flour-improving and meat-curing agent.

Furthermore, the fact that vitamin C is found in all living tissues, both animal and vegetable, and that its presence in many fruits and vegetables is relatively large, makes the compound physiologically acceptable and safe. Large scale manufacture, coupled with high standards of purity and relatively low cost, also make applications economically feasible.

Finally, I should like to mention a very important factor: the world-wide legal acceptance of vitamin C and many of its technical uses. General admittance is not only based on the fact that vitamin C is present in most natural foods, but also on the results of toxicological studies which indicate the safety of human intakes of up to 4 g/day.[3,4] A list of fields of application and the corresponding amounts of vitamin C proposed for use in the EEC is reproduced below (Table 1).

The most important fields of application will be dealt with in detail in other papers today and tomorrow and I will not, and cannot, anticipate these lectures, but I should like to make a few introductory and perhaps complementary remarks mainly based on recent work.

TABLE 1

Technical applications of vitamin C levels proposed in the EEC list

Fruit juices and fruit juice beverages, lemonades, syrups	200 mg/litre
Beer	100 mg/litre
Wine	200 mg/litre
Meat and meat products (curing)	1 g/kg
Fresh meat products	2 g/kg
Frozen fruits and vegetables	2 g/kg
Canned fruit	2 g/kg
Flour or bread improver	100 mg/kg
Fish, frozen or processed	1 g/kg
Margarine	300 mg/kg
Milk, dry, non-fat	500 mg/kg
Yoghourt	500 mg/kg
Potatoes raw, peeled	2 g/kg
Potato products	2 g/kg

Meat Curing

A real problem of paramount interest is that of meat curing and health, that is, the formation of nitrosamines and their potential carcinogenic effect. The world-wide anxiety about harmful additives and residues in foods has led in some countries, like Norway and the USSR, to restrictions regarding the use of nitrites and nitrates when their potential danger to health—as agents for the formation of toxic nitrosamines—was recognised.

However, the problem is extremely complex and many factors require detailed examination, such as:

the conditions under which nitrosamines are formed in meat products;

the possibilities of reducing nitrate and nitrite, both in the amount added as well as the residual quantity;

the effects on organoleptical properties of such a reduction; and, most important also, the microbiological point of view, with particular reference to pathogenic and toxinogenic bacteria.

The main object of all these studies is, of course, the complete elimination of any toxic hazards which may be connected with the use of nitrites. Recent investigations indicate that ascorbate reduces the formation of nitrosamines, and particularly of N-nitrosopyrrolidine, during the frying of bacon; the problem, however, is that very high amounts of vitamin C appear to be necessary. The results of a first study in the USA with 1000 ppm ascorbate were 'encouraging, but inconclusive', and tests with 2000 ppm were proposed in September 1973 by the American Meat Institute.[5] I am, therefore, looking forward to Dr Walters' lecture for further results in this field.

Antioxidants other than ascorbate have been discussed as possible nitrosamine inhibitors, but ascorbate has the advantage of being 'unquestionably safe' as USDA officials have pointed out.[6] The determining factor when considering a reduction of nitrites is their bacteriological influence on meat products, and so far no substitute for nitrite has been found which not only inhibits botulism, but also favourably affects meat colour, flavour and texture.

The formation of carcinogenic N-nitroso compounds by the chemical reaction between nitrous acid and various amines—including drugs—is blocked by ascorbic acid. Mirvish *et al.*[7] found that the extent of blocking, the reaction rate and the type of chemical reactions occurring depended on the experimental conditions. Nevertheless these authors have suggested that drugs which can be N-nitrosated could be combined with ascorbate before they are administered.

Ascorbic Acid as a Flour Improver

As we can see from the list of lectures at this Symposium, there are other fields of application where vitamin C has its functional uses. The Chorleywood continuous bread making process uses ascorbic acid as an improver in a higher than usual proportion, and because of the broad use of this successful process—which represents a significant advance in food technology—the fate of ascorbic acid in the process has been carefully studied.[8]

It is assumed that the flour improvement is actually based on dehydro-ascorbic acid, *i.e.* that ascorbic acid has to be oxidised first in order to react. In contrast to this particular mechanism, most of the technical uses of ascorbic acid are based on its antioxidative and general stabilising effect.

Canned and Bottled Foods

Experience, backed by theory, shows that a sufficient amount of undecomposed ascorbic acid must be present in bottled, canned, or otherwise hermetically closed containers in order to achieve good results. That means, quite generally, that the amount of ascorbic acid added must suffice to guarantee that a sufficient part survives processing, filling and subsequent storage and—depending on the type of product—reconstitution, cooking, and so on. It also leads to a general conclusion that processing and storage conditions must be chosen in such a way that losses of ascorbic acid are minimised; this means that the precautions which are to be observed when foodstuffs containing ascorbic acid are processed—whether the ascorbic acid is added or is naturally present—should ideally be those ensuring optimal vitamin C stability. It is essential, for practical success, to fulfil as strictly as possible the following precautions:

(1) The equipment used should be of stainless steel, aluminium, enamel, glass, china or plastic.
(2) Direct contact of the food product or its ingredients with brass, bronze, copper and iron must be avoided.
(3) Wherever possible de-aeration should precede processing, which should be carried out under inert gas or in a vacuum.
(4) During mixing, emulsification, homogenisation, etc., no air should be introduced into the product.
(5) Containers should be filled to maximum capacity, *i.e.* the head-space should be kept as small as possible.
(6) Whenever feasible, a short-time heat treatment should be employed to inactivate enzymes.
(7) Micro-organisms should be removed by filtration or inactivated by heat treatment, if possible, and processing should continue under aseptic conditions.
(8) The products should be stored at low temperature.
(9) The material must be protected from light and other radiant energy.
(10) If practicable, sequestering agents such as phosphates, citrates, EDTA, or cysteine may be added.

(11) Preferably, some of the ascorbic acid should be added just prior to bottling or canning, etc. (1 ml of residual air requires theoretically 3·3 mg of ascorbic acid).

(12) All autoxidisable ingredients, such as flavouring oils, should have a low peroxide value.

In evaluating the stabilising effect of added ascorbic acid I should like to stress a point which is of general validity in food manufacture: although chemical and physical parameters are good indicators of the quality of a product and may be helpful in screening tests, in the final analysis the factors that really matter are taste and appearance.

All studies dealing with the stabilisation of ascorbic acid are, therefore, potentially interesting, particularly in view of the technical application of this vitamin. Many works deal with the aerobic and anaerobic degradation of ascorbic acid, and in a recent publication, Herrmann[9] reviews the protective effect of flavonoids and other phenolic compounds largely occurring in nature. The antioxidative properties of vegetables and spices may often depend on plant phenolics.[10]

As a possible means of maintaining the ascorbic acid content of orange juice during storage, the removal of oxygen by the addition of glucose oxidase has been examined. An antioxidant effect was obtained only when catalase was added simultaneously; with glucose oxidase alone the loss of ascorbic acid was greater than in juice with no enzyme addition.[11]

After the disappearance of free oxygen from stored canned foods, usually within a month of sealing, subsequent loss of ascorbic acid is due to anaerobic decomposition[12,13] which yields carbon dioxide and 5-carbon compounds. The reaction which predominates below pH 2 gives furfural as the major product. The reactions occurring in the pH range of foods, *i.e.* the reaction with an optimum near the pH of ascorbic acid and the fructose-promoted reaction may give 2,5-dihydro-2-furoic acid as the major product, but this has not yet been established. Quantitative determinations of this acid during the decomposition of ascorbic acid over a range of pH would be necessary to answer the question.

Antioxidant Mixtures

A potentially very interesting application is the use of fat-soluble vitamin C esters, such as ascorbyl palmitate, in antioxidant mixtures for fats and oils.[10,14] The effectiveness of synergistic mixtures of α-tocopherol, ascorbyl palmitate and lecithin, and particularly of those containing, in addition, a gallate, is very good and higher than that of BHA and BHT. Some results published recently by Pongracz[15] are reproduced in the Tables 2 and 3, and Figs. 1 and 2, and they show clearly the enhanced efficacy of mixtures containing ascorbyl palmitate as an antioxidant.

It is a pity, however, that in spite of all these good results, ascorbyl palmitate has such a poor solubility and slow dissolution rate that this represents a significant disadvantage and drawback for many practical applications. Efforts to overcome these difficulties by using special mixtures with partial glycerides or high amounts of phosphatides have been unsuccessful, for one reason or another. Theoretically a solution would be offered by the following: the replacement of erythorbyl palmitate for ascorbyl palmitate. Equal antioxidative effectiveness and higher solubility are the advantages, but the limited legal acceptability of erythorbyl palmitate as a food additive,

TABLE 2

Stability of butterfat stored at 80°C

Antioxidant in ppm	Peroxide-value after x days			
	3	4	5	6
Control	265	—	—	—
500 AP	74	280	—	—
200α–TL	9·9	56·3	110	
100 BHT + 100 BHA	5·3	47·6	100	
500 AP + 200α–TL	1·3	3·7	4·4	11·0
500 AP + 100 OG	2·0	2·5	3·0	3·5
500 AP + 200α–TL + 100 OG	1·1	2·1	3·2	4·6
500 AP + 500 LAF + 100α–TL (A)	0·9	1·5	2·8	4·5
500 AP + 500 LAF + 100α–TL + 100 OG (G)	1·0	1·6	2·0	2·5
500 AP + 500 LAF + 100α–TL	0·5	1·0	1·2	1·5

AP: Ascorbyl palmitate
TL: Tocopherol
OG: Octyl gallate
LAF: Lecithin-fraction

TABLE 3

Stability of sunflower oil at 80°C

Antioxidant in ppm	Peroxide-value	
	2 days	3 days
500 AP + 500 LAF + 100α–TL (A)	0	1·8
300 AP + 300 LAF + 60α–TL (A)	0	38·4
500 AP + 500 LAF + 100α–TL + 100 OG (G)	0	1·6
300 AP + 300 LAF + 60α–TL + 60 OG (G)	0	22·1
100 BHT + 100 BHA	165	>400
Control	200	>400

coupled with lack of vitamin C activity, are the negative sides. The reduced uptake of ascorbic acid after the administration of erythorbic acid (Hornig[15a]) is not expected to play a significant role here, since the amount of erythorbic acid absorbed would be limited in two ways: first, the amount needed as an antioxidant would be restricted to approximately 500 mg/kg of fat or oil and, secondly, the amount of stabilised oil consumed would usually not exceed 50 g, *i.e.* the daily consumption of erythorbate would amount to 25 mg corresponding to approximately 10 mg of erythorbic acid. Considering

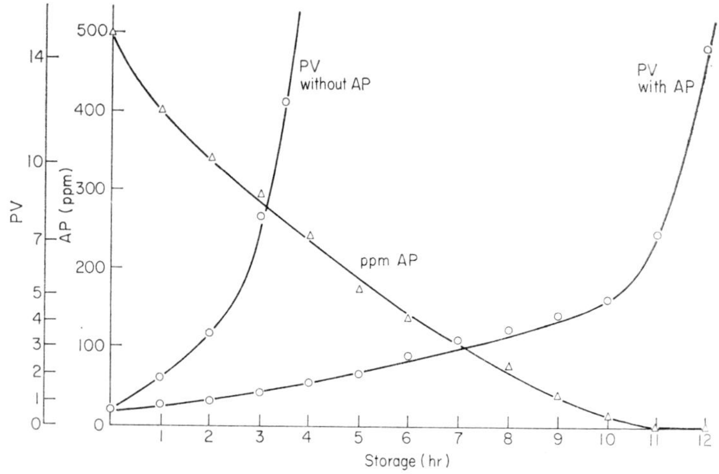

FIG. 1. Antioxidant test in sunflower oil at 100°C. Correlation between amount of AP and stability of sunflower oil.

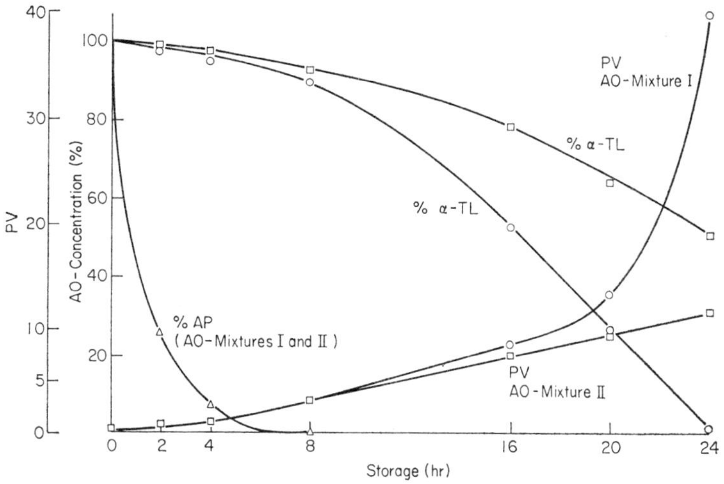

FIG. 2. Antioxidant test in lard at 100°C. Correlation between amount of antioxidant and stability of lard. AO – mixture I = 500 ppm AP + 100 ppm α–TL; AO – mixture II = 500 ppm AP + 100 ppm α–TL.

the slow and probably incomplete absorption from the large quantity of oil, we can safely assume that this small amount of 10 mg would not competitively reduce the uptake of vitamin C.

Vitamin C Enrichment

Another practical problem has arisen in the last few years in connection with the vitamin C enrichment of breakfast beverage powders, especially baby foods. A comparative study including ascorbic acid, sodium ascorbate and ascorbyl palmitate in various mixtures shows a significant influence of the type of carrier used on the retention values[16] (Table 4).

Ascorbic acid in cocoa-based breakfast beverage powders is best added in the form of the fat-coated product, which has proved to be more stable in these types of foods. The crystals of ascorbic acid are coated with a saturated triglyceride, and this coating slightly delays dissolution in water but significantly improves stability (*see* Figs. 3 and 4).

Segregation of added vitamins may lead to difficulties when the dry food product consists of a relatively coarse material, *e.g.* flakes or granules, granulated baby food, extruded snacks and the like. In

TABLE 4

Stability of various vitamin C forms in some edible carriers (after one year storage in closed bottles)

Vitamin added	Temperature	Vitamin C-retention in % of initial value				
		Wheat starch	Wheat flour	Skimmed milk powder	Glucose monohydrate	Glucose anhydrous
Ascorbic acid	RT	97·8	71·2	70·7	76·8	97·2
	45°C	83·7	68·5	70·2	74·2	75·8
Sodium ascorbate	RT	97·5	70·4	81·3	84·5	94·8
	45°C	81·4	70·2	72·1	74·5	74·2
Ascorbyl palmitate	RT	95·8	69·3	86·0	90·1	85·9
	45°C	85·4	43·4	81·8	55·3	80·4
Water content of carrier material		12·0%	12·5%	3·5%	9·1%	0·1%

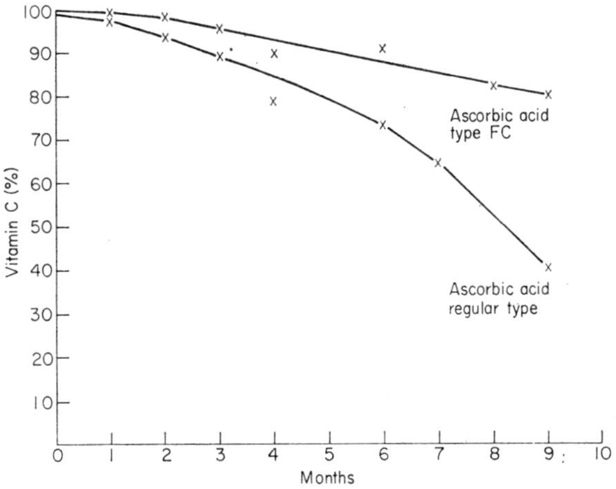

FIG. 3. Stability of vitamin C in breakfast beverage powder. Addition of vitamin C *after* instantising.

particular, the very fine vitamin powders will segregate off smooth, flat surfaces having low powder-carrying capacities, and it will not be possible to achieve a homogeneous distribution. In cases where vitamin addition to the dough or paste before drying is not feasible because of destruction during subsequent processing, *e.g.* during an

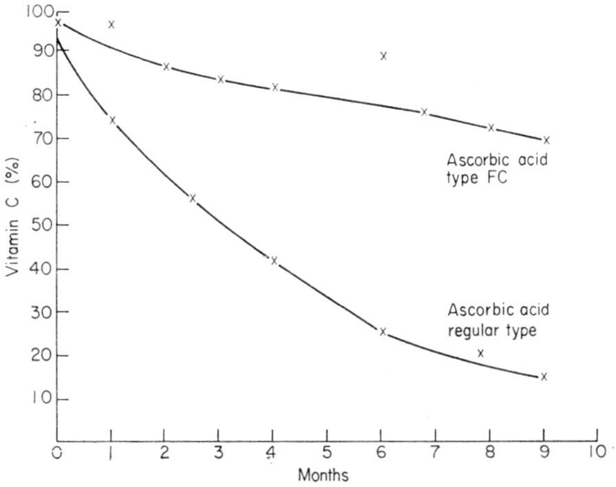

FIG. 4. Stability of vitamin C in breakfast beverage powder. Addition of vitamin C *before* instantising.

extra heat treatment, special procedures applying the vitamins in a suspension of molten fat or oil have been developed in addition to the conventional methods of adding in syrups.[17,18]

Abscission of Fruits

I should like to mention a technological application of ascorbic acid which might be considered as a borderline food use: the induction of the abscission of fruits by spraying with an ascorbic acid solution.

Tests were carried out in California in 1966 and 1967 by Wilson and Hartmann with different varieties of oranges and olive fruits. The main object was to find chemicals which would give good activity with a minimum of tree and fruit injury, since these factors would greatly improve the economics of mechanical harvesting.[19,20] When olives were sprayed under relatively dry conditions ascorbic acid was not effective in stimulating fruit abscission. However, when applications preceded or followed periods of rainy weather a definite reduction in the required removal force was achieved.

The work is still in the experimental stage. It is postulated that a possible mechanism is the interaction of ascorbic acid with auxin. It has long been known that abscission of plant parts, including fruits and leaves, is a correlation effect influenced by auxin, and ascorbic acid is considered to be a plant growth hormone behaving as an auxin antagonist. The antagonistic effect may be brought about by a direct interaction between the native auxin and the oxidation–reduction state of the ascorbic acid system in the regulation of growth.

Another mechanism is being considered for oranges, where ethylene gas seems to regulate abscission besides having other effects, such as on ripening. The idea of trying ascorbic acid for oranges was suggested by Hartmann's work which demonstrated that ascorbic acid was an abscission agent for olives. The results with oranges, however, do not look very promising, since no effect was observed with most of the chemicals tested.

To conclude, I would like to explain that it would be an impossible task to deal with all the known applications of vitamin C as a chemical. To do so adequately would take more time than the whole of this Symposium. Thus, I have concentrated on a few uses which I feel are more important and more interesting. I hope that they will have served to indicate the scope of the subject.

REFERENCES

1. Ammon, R. (1964). *Wiss. Veroff. Dtsch. Ges. Ernahrg.*, **14,** p. 206.
2. Kläui, H. (1970). *The Functional (Technical) Uses of Vitamins,* University of Nottingham Seminar on Vitamins, Ed. Mendel Stein, Churchill Livingstone, Edinburgh and London, p. 110.
3. Hanck, A. (1974). *Int. J. Vit. Nutr. Res.* (in press).
4. Korner, W. F. and Weber, F. (1972). *Int. J. Vit. Nutr. Res.*, **42,** p. 528.
5. American Meat Institute Report (1973). *Food Chem. News,* **15,** p. 24.
6. USDA Report (1973). *Ibid.*, **15,** p. 12.
7. Mirvish, S. S., Wallcave, L., Eagen, M. and Shubik, P. (1972). *Science,* **177,** p. 65.
8. Thewlis, B. H. (1971). *J. Sci. Fd Agric.*, **22,** p. 16.
9. Herrmann, K. (1973). *Fette Seifen Anstrichm.*, **75,** p. 499.
10. Kläui, H. (1973). *Naturally Occurring Antioxidants,* Symposium on Rancidity in Fatty Foods, University of Technology, Loughborough. *IFST Proceedings,* **6,** p. 195.
11. Tschogowads, S. K. (1972). *Lebensmittel-Industrie,* **19,** p. 287.
12. Finholt, P. *et al.* (1965). *J. Pharm. Sci.*, **54,** p. 181.
13. Huelin, F. E. *et al.* (1971). *J. Sci. Fd Agric.*, **22,** p. 540.
14. Kläui, H. (1972). *Anwendung der Vitamine C und E als Antioxydantien in der Lebensmitteltechnologie,* CIIA Congress, Saarbrücken.
15. Pongracz, G. (1973). *Int. J. Vit. Nutr. Res.*, **43**(4), p. 517.

15a. Hornig, D. (1974). *Experientia* (in press).

16. Pongracz, G. (1972). *Unpublished data,* F. Hoffmann-La Roche & Co. Ltd, Basle, Switzerland.
17. F. Hoffmann-La Roche & Co. Ltd, Basle. *Pat. appl. filed.*
18. Quaker Oats USA, *Brit. Pat.* 1 327 350.
19. Wilson, W. C. *Progress on Citrus Fruit Abscission During the* 1966–1967 *Fruit Season.* Florida Citrus Commission, Lake Alfred, Florida.
20. Hartmann, H. T. *et al.* (1967). *Calif. Agr.*, **21,** p. 5.

DISCUSSION

Wyeth: Can Dr Kläui give us any information about the use and properties of potassium ascorbyl palmitate or sodium ascorbyl palmitate?

Kläui: These are compounds that behave in a manner similar to soaps. Of course they are very sensitive to calcium because this will lead to a precipitate, but their surface-active properties make them quite interesting since they will disperse much more easily than the free ascorbyl palmitate, especially in multi-phase systems.

Pawan: May I ask if the proposed EEC list is to be discussed this evening? I wondered how the values you gave—those in the first table you showed—compared with present values in most European countries.

Chairman: I would have thought it probably would be discussed in one of the study groups, yes. It seems to be one of the more important aspects of the EEC legislation.
Coultate: Could you tell us whether palmityl ascorbate has similar biological activity to the parent vitamin?
Kläui: Corresponding to molecular weight, it has full biological activity.

3

The Chemical Estimation of Vitamin C

J. R. COOKE

Laboratory of the Government Chemist,
Cornwall House, Stamford Street, London, England

ABSTRACT

The extraction of vitamin C from foodstuffs and other samples is discussed with particular reference to the estimation of the percentage recovery of the vitamin in a multiple extraction procedure. The methods available for the estimation of the vitamin C content of sample extracts are reviewed, the procedures being grouped according to which of the two reactive compounds, ascorbic acid or dehydroascorbic acid, are being measured.

Some remarks are made concerning the author's experience of several of the methods and certain modifications to one of the methods are discussed. Finally, a few suggestions are made regarding the prospects for new and better methods for the estimation of vitamin C activity.

INTRODUCTION

The chemical estimation of vitamin C is fundamentally a two-part process. The vitamin has to be extracted from the sample and then its concentration has to be estimated.

EXTRACTION

Several papers have been published[1–3] comparing different extraction solvents but these papers have failed to arrive at a unanimous conclusion. The preferred solutions are those of metaphosphoric or oxalic acid in water with or without the addition of other reagents, *e.g.* EDTA and acetic acid. A solution consisting of 3% metaphosphoric acid and 8% acetic acid is used at this Laboratory for the extraction of vitamin C from foodstuffs and sometimes, *e.g.* with cooked potato, 25% of acetone is incorporated in order to obtain a satisfactory separation of solid and liquid phases. These two solutions precipitate protein, stabilise the vitamin and prevent absorption of the vitamin by the active charcoal which is used in at least one method of estimation. Papers frequently omit details of the extraction procedure and errors can be introduced if a satisfactory method is not employed. If a small amount of solid material, say 1 g, is treated once with 100 ml of solution, it is possible to extract almost all of the vitamin and the volume of liquid held by the solid sample can be ignored. However, with foodstuffs it is usually necessary to extract, say, 50 g of a sample with the minimum volume of extracting solution in order to obtain a reasonable concentration of the vitamin. In such circumstances the solid material can retain up to a quarter of the extracting solution and a similar proportion of the vitamin even after the extract homogenate has been centrifuged. The solid material must be extracted as many times as are necessary to obtain practically all of the vitamin or, alternatively, the concentration of the vitamin in the extract or extracts must be corrected for loss.

In a scheme of sequential extractions, it can be shown theoretically that a series of small extractions should obtain the vitamin in a minimum volume of solution, whereas a smaller number of large extractions will give either a larger volume of solution or a lower recovery. In practice, this does not always seem to be so. In two cases studied, the maximum recovery was obtained when a volume of extracting solution was divided into one large extract followed by one or two small extracts. The reason probably lies in the losses of vitamin C which can occur when low concentrations of the vitamin are in the presence of freshly macerated material. It appears to be best to obtain as much as possible of the vitamin in the first extract and this entails the use of a large volume of extracting solution. We usually carry out one large extract, say 250 ml for 50 g of sample, followed by one smaller extract, say 100 ml, and multiply the amount of vitamin found by a calculated correction factor.

The correction factor is calculated from the volumes of sample, extracting solution and extract recovered. In a simple case the percentage recovery in the first extraction will be

$$\frac{V_{1\ \mathrm{recovered}}}{V_{\mathrm{sample}} + V_{1\ \mathrm{extractant}}} \times 100$$

and in the second extraction, the recovery expressed as a percentage of the vitamin remaining in the sample will be

$$\frac{V_{2\ \text{recovered}}}{V_{\text{sample}} + V_{1\ \text{extractant}} - V_{1\ \text{recovered}} + V_{2\ \text{extractant}}} \times 100$$

For foodstuffs containing a high proportion of water, the volume, if not measurable, is obtained from the weight of sample by assuming a value of unity for the density.

The total percentage recovery from the two extracts is used to correct the total amount of vitamin found. Calculations can be extended to cover further extractions or to allow for the presence of appreciable amounts of non-soluble matter. The proportions of the vitamin found in each of a series of extractions are close to the calculated levels and even the simple procedure gives a fair estimate of the recovery of vitamin from a sample.

ESTIMATION

The estimation of the amounts of vitamin C in sample extracts has been the subject of a large number of scientific papers. This is perhaps an indication that workers in this field are not altogether satisfied with the analytical methods so far proposed.

Two naturally occurring compounds have been found to have antiscorbutic activity, ascorbic acid (AA) and its oxidation product dehydro-ascorbic acid (DHAA) which has 80–100% of the activity of AA. Diketogulonic acid (DKGA) which is the further breakdown product of dehydro-ascorbic acid is not biologically active and for all practical purposes its formation can be considered an irreversible reaction.

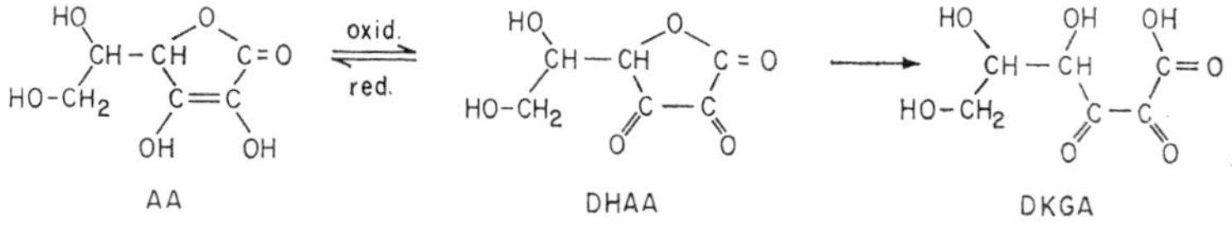

Hence the problem in measuring vitamin C activity by chemical means is to determine the amounts of AA and DHAA in a sample and to avoid interference from any DKGA or other non-active substances. In most foodstuffs the levels of DHAA and DKGA are low and an approximation to the level of vitamin C activity can be obtained by estimating the level of AA alone but for the most accurate work this is not satisfactory. For analytical ease and because of the easy interconvertibility of AA and DHAA, it is usual to measure the concentration of one compound together with the sum of the two on another aliquot of extract after an oxidation or reduction stage.

Estimation of Ascorbic Acid

The types of reactions which have been used for determining AA are very limited. Apart from the oxidation of AA to DHAA using a wide variety of oxidising agents, only the reaction with diazotised nitroaniline and similar compounds has been utilised. Some of the most popular methods for the estimation of ascorbic acid are those based on its reducing power. Reagents which have been used include 2,6-dichlorophenolindophenol, 2,2′-bipyridyl, N-bromosuccinimide, iodine, mercuric chloride and many others. In addition, polarographic, potentiometric and enzymic methods have a similar basis. All these oxidation procedures are likely to suffer from interference by other reducing agents which might be present in the extracts. Many variants of the methods have been published, most giving details of procedures designed to limit the effect of reducing substances other than ascorbic acid, such substances being particularly troublesome in the case of cooked foods. Of all these methods the most popular by far is the titration with 2,6-dichlorophenolindophenol.[4] The method suffers from interferences and there are problems in seeing the end-point of the titration when sample extracts are coloured. However, despite these shortcomings, the method gives rapid and satisfactory results for many commodities.

Enzyme methods using ascorbic oxidase have not been widely used but several versions of reagents based on diazotised nitroanilines[5] have been reported. The reaction is complicated and the ascorbic acid molecule is partly destroyed in forming the coloured product. One of the most elaborate versions of this method is that of Crossland[6] where a chromatographic separation stage precedes the colour development reaction. We have examined a slightly modified version of this method and after overcoming some problems we achieved reasonably satisfactory results. The procedure is time consuming and there is no provision for reducing any DHAA in the sample extract to AA and hence, any DHAA will not be included in the result. However, some other procedures incorporate a stage to reduce DHAA to AA and there seems no reason why this should not be done in this method. Practice is required if AA is not to be lost, particularly during the concentration of the extract solution prior to transfer to a cellulose chromatographic column. It is essential to carry out much of the procedure in the absence of day-light or large losses of AA occur during the concentration and chromatographic stages.

Estimation of Dehydro-ascorbic Acid

Use has been made of at least three reactions of DHAA for the estimation of vitamin C. AA is oxidised to DHAA prior to the estimation and these methods give an estimate of total vitamin C activity. The three methods are based on reactions with 2,4-dinitro-phenylhydrazine, ortho-phenylenediamine and glycine. The method based on the reaction with glycine[7] has not been widely used and it has not been refined to the same extent as the other two methods.

The reaction with 2,4-dinitrophenylhydrazine[8] is very well known and in its most sophisticated forms is probably the most accurate method for the estimation of AA and DHAA, in a wide range of products. This procedure has been adopted by the EEC for the estimation of vitamin C in foodstuffs. The method is very time consuming if the most accurate procedure, which contains two chro-matographic separations, is used. It is possible to extend the method even further, when values for DKGA, DHAA plus DKGA and AA plus DHAA plus DKGA can be obtained. In the more usual procedure any DKGA present is estimated along with the total AA plus DHAA.

The fluorimetric method based on the reaction between DHAA and ortho-phenylenediamine was developed by Deutsch and Weeks[9] and is now an AOAC procedure.[10] The method will eliminate the effects of most interfering substances yet is not too time consuming. Results represent the total AA plus DHAA but DKGA has been said to interfere in the reaction with ortho-phenylenediamine.[11] The procedure is based on the production of a fluorescent quinoxaline derivative from the condensation of DHAA and ortho-phenylene-diamine. It is quite possible that substances other than DHAA could form compounds which fluoresce under the test conditions and these would interfere. This is the type of problem which is common to most methods for the estimation of vitamin C. In this method, as in some others, there is an attempt to eliminate the effect of most, and hopefully all, interfering substances by introducing another obstacle which substances must overcome if they are to interfere. This obstacle is in the form of a parallel determination which gives rise to fluorescence in the usual manner but only by those compounds which will not couple with boric acid. The conditions for the coupling reaction with boric acid are those under which virtually all of the DHAA does couple. From readings for the total fluorescence and for that given only by those compounds which do not couple with boric acid, it is possible to obtain the fluorescence corresponding to those compounds which do couple with boric acid. It is hoped that only DHAA will be measured in this way.

We have looked at several aspects of this method. As published, the reaction between DHAA and boric acid is allowed to proceed for 15 min but we find that the reaction is not complete for an hour or more and that a half hour is needed for the reaction to be practically complete. The lowest level of vitamin C which can be measured by the method is governed by the fluorimeter sensitivity, the scale of the calibration curve and the proportion of the total fluorescence which is produced by substances other than DHAA. The level of such substances is low in most foodstuffs and hence the method can be adapted to measure lower levels of vitamin C than is possible using the Deutsch and Weeks procedure, provided a sufficiently sensitive fluorimeter is available. Using such a modified procedure it has been possible to measure the levels of vitamin C in milk, both fresh and boiled, up to three days after purchase. Vitamin C levels as low as 0·1 mg/100 g of sample can be measured. Where the vitamin C level is unknown, two calibration curves are prepared covering a very wide range of levels of the vitamin. When the fluorescence of the sample is measured the instrument sensitivity is set to the value used for whichever calibration curve is appropriate. It is also possible to obtain an estimate of the level of the vitamin by titrating an aliquot of the sample extract with 2,6-dichlorophenol-indophenol prior to carrying out the fluorimetric procedure. We have found this fluorimetric method is very versatile and we have used it for a wide range of foodstuffs including liver, milk, fruit, tinned fruit, vegetables both raw and cooked, potato powder and a variety of other products.

FUTURE DEVELOPMENTS

The 2,6-dichlorophenolindophenol and 2,4-dinitrophenylhydrazine methods have been in use for some 40 and 30 years respectively although refinements to these procedures are still being published. Most of the other methods mentioned have been developed more recently but even here up to 10 years have elapsed without any really new developments being reported. In the last 10 years analytical chemistry has changed out of all recognition and it seems most unlikely that developments in the estimation of vitamin C activity will not be reported in the near future.

Indeed some developments have already taken place, but these have not so far resulted in a new method for use with foodstuffs and other complex samples. In 1963 Sweeley *et al.*[12] reported the gas chromatography of silylated ascorbic acid. Similar reports by other workers soon followed and in 1968 Pfeilsticker[13] successfully separated the silylated forms of AA and DHAA. Everything seemed set for a gas chromatographic method for the estimation of vitamin C but so far this has not occurred. Allison and Stewart[14] have reported the estimation of AA in the brains of rats by GLC but the small size of the samples made the case atypical. It seems more than likely that the difficulties are connected with obtaining extracts of samples in a suitable form for silylation. Before silylation can be carried out a sample extract will need to be almost anhydrous or the reagent will react with any water present. Such a requirement almost certainly precludes the use of the usual aqueous acid extracting solutions. Perusal of the literature on the estimation of vitamin C reveals the use of only four extracting solutions not primarily based on aqueous extracting mixtures. The solutions were methanol,[15] methanol–light petroleum–aqueous oxalic acid,[16] 70% ethanol containing 2-mercaptoethanol,[17] and benzene-dimethyl formamide containing succinic acid.[18]

Little information is available on the last solution but the other three have been reported to be satisfactory, so there does seem to be a prospect of extracting vitamin C from foodstuffs using a methanol or ethanol based extracting solution containing a small proportion of water at the most. It should be possible to concentrate such extracts on a rotary evaporator at no higher than room temperature removing any water at the same time. The remaining problem would be to get the vitamin C dissolved in a solvent which is compatible with the silylating agent yet would not upset the running of the GLC column.

Over the years several papers have appeared giving details of enzymatic methods for the estimation of vitamin C using ascorbic acid oxidase.[15,19,20] Unfortunately the enzymes have never been specific for ascorbic acid. In other fields there have been big advances in the use of enzymes for analytical purposes and it seems probable that improvements in enzymatic methods for vitamin C could be on the way.

There is a rapidly expanding area of analytical chemistry where as yet no publications referring to vitamin C have appeared and that is in high pressure liquid chromatography (LLC). There seems to be no reason why AA and DHAA cannot be separated from other substances and from each other by LLC. As it should not be necessary to prepare a derivative, the normal extraction procedure would probably be applicable. Column, paper and thin layer chromatography have already been used in separations of AA and DHAA and in separations of the vitamin from other substances.[6,18,21] Although such techniques have been moderately successful the much greater separation efficiencies of both LLC and GLC will be valuable in ensuring that the vitamin is separated from all other substances. In addition, LLC and GLC procedures do not require a chemical estimation stage after the separation of the vitamin.

If any one of the suggested procedures should be developed we may at last have available a method giving rise to unequivocal results for the vitamin C contents of samples.

ACKNOWLEDGEMENT

The author wishes to thank the Government Chemist for permission to publish this paper.

REFERENCES

1. Feldheim, W. and Seidemann, J. (1962). *Fruchtsaft Ind.*, **7**, p. 166.
2. Istratescu, L. (1964). *Farmacia* (*Bucharest*), **12**(6), p. 349.
3. Pribela, A. and Pikulikova, C. (1969). *Prum. Potravin*, **20**(1), p. 27.
4. Tillmans, J., Hirsch, P. and Hirsch, W. (1932). *Z. Untersuch. Lebensm.*, **63**, p. 1.

5. Schmall, M., Pifer, C. W. and Wollish, E. G. (1953). *Analyt. Chem.*, **25,** p. 1486.
6. Crossland, I. (1960). *Acta Chem. Scand.*, **14**(4), p. 805.
7. Brunet, M. (1968). *Ann. Pharm., Fr.*, **26**(12), p. 797.
8. Roe, J. H. and Kuether, C. A. (1943). *J. Biol. Chem.*, **147,** p. 399.
9. Deutsch, M. J. and Weeks, C. E. (1965). *J. Assoc. Off. Agric. Chem.*, **48**(6), p. 1248.
10. *Official Methods of Analysis of the Association of Official Analytical Chemists,* 11th edn, p. 778, AOAC, Washington D.C. (1970).
11. Davidek, J., Velisck, J. and Nezbedova, M. (1971). *Sb. Vys. Sk. Chem.-Technol. Praze Potraviny,* **E30,** p. 17.
12. Sweeley, C. C., Bentley, R., Makita, M. and Wells, W. (1963). *J. Amer. Chem. Soc.*, **85,** p. 497.
13. Pfeilsticker, K. (1968). *Fresenius' Z. Anal. Chem.*, **237**(2), p. 97.
14. Allison, J. H. and Stewart, M. A. (1971). *Analyt. Biochem.*, **43,** p. 401.
15. Marchesini, A., Polesselo, A. and Zoja, G. (1970). *Agrochimica,* **14,** p. 453.
16. Müller-Mulot, W. (1964). *Deut. Apotheker-Ztg.*, **104,** p. 469.
17. Friberg, U., Lohmander, S. and Carlsson, G. (1970). *Arkiv. fur Kemi,* **31**(38), p. 467.
18. Müller-Mulot, W., unpublished, *see* Ref. 19.
19. De Ritter, E. (1965). *J. Assoc. Off. Agric. Chem.*, **48**(5), p. 985.
20. Marchesini, A. and Manitto, P. (1972). *Agrochimica,* **16,** p. 351.
21. Hegenaur, J. and Saltman, P. (1972). *J. Chromatog.*, **74**(1), p. 133.

DISCUSSION

Hornig: I should like to hear your opinion on the differentiation between ascorbic acid and isoascorbic acid.

Cooke: There has been a paper published on this, an American paper by Deutsch and Weeks, in which they carried out a separation by chromatography.

Hornig: Qualitative method or quantitative?

Cooke: I think it was reasonably quantitative. It was a good time ago, about 1965, or maybe a little earlier than that, in JAOC.

Walters: Are there any stable preparations of oxidising enzymes which could be useful in spite of their non-specificity?

Cooke: This is a little difficult because the people working in this field are Italians. It looks as if there might be something coming along, but it is difficult, at least as far as I'm concerned, to get translations from the Italian, and to see how much there is there. Three or four papers have been published in the last three years—the name, I think, was Marchetti.

4

Quality Changes Related to Vitamin C in Fruit Juice and Vegetable Processing

G. G. Birch, B. M. Bointon, E. J. Rolfe and J. D. Selman

National College of Food Technology,
University of Reading, Weybridge, Surrey, England

ABSTRACT

Although vitamin C is well known both as an indicator and as an agent of quality in food processing, the mechanism of its action is by no means understood. This paper presents some preliminary observations in a twofold project, designed to explore the basic significance of vitamin C concentration, during the processing of fruit juice and vegetables. Using model orange juice systems it can be established that glucose and fructose of the three major sugar components contribute to pasteurisation off-flavour, but it is probable that the volatile oil component of whole juice is of much greater importance.

An indication of the extent of loss of vitamin C from peas during water blanching, due to leaching, oxidation, and thermal degradation, has been obtained by measuring the ascorbic acid (AA) and dehydro-ascorbic acid (DHA) in whole peas, cotyledons, testa, and the blanch water before and after blanching for different times and temperatures.

INTRODUCTION

It is recognised that fruit juices undergo a flavour change on pasteurisation, the magnitude of the change varying from juice to juice and from process to process. The effects of ascorbic acid and oxygen were investigated, in the current project, and an attempt was made to discover whether the aqueous phase is involved in the production of pasteurisation off-flavour. In preparing aqueous model fruit juice systems, it was noted that panellists, in significant numbers, could differentiate between processed and unprocessed fructose and glucose solutions, but not processed and unprocessed sucrose solutions. Aspects of this flavour change were investigated.

The loss of vitamin C from peas during water blanching has been studied by several workers over the last forty years and in many cases considerable losses have been reported, the results indicating that the loss was primarily due to leaching. However, a more detailed review of the literature revealed that a wide range of blanching times and temperatures had been studied using different pea varieties and different types of commercial equipment. In most cases at least one important factor, which might have affected the loss of vitamin C, had been ignored so any comparative analyses of the results were virtually impossible. Thus vitamin C losses ranging from 3 to 30% have been reported for a one minute blanch at 95–100°C. Since several factors may affect both the vitamin C content at the time of blanching, and the losses incurred during blanching, blanching operations were carried out on a laboratory scale in order to obtain a detailed picture of the location and state of the vitamin C, during blanching, as a basis for further work.

QUALITY CHANGES DURING FRUIT JUICE PROCESSING

Experimental Methods

(*a*) *Fruit juices.* The fruit juices used were reconstituted frozen orange concentrate (ex Israel) and reconstituted hot packed pineapple concentrate (ex Philippines). The juices were aerated by passing air for 15 min or deoxygenated by passing nitrogen until the oxygen level was less than 2 ppm (oxygen meter) before pasteurising by heating to 90°C and bottling. The bottles were immediately cooled and refrigerated (5°C) overnight.

When the level of ascorbic acid was raised, the required amount of Analar (BDH) L-ascorbic acid was added.

De-oiled orange juice was prepared by extracting reconstituted orange juice with three volumes of diethyl ether. The aqueous phase was separated and a trace of silicone anti-foam added. Nitrogen was then passed through the orange juice until all trace of diethyl ether had disappeared. Half of this orange juice was processed as above and both halves were refrigerated overnight.

(*b*) *Model systems.* Sugar solutions were prepared from Analar (BDH) sugars. The solutions, in 2 litre batches, were aerated by bubbling air for 15 min after the addition of ascorbic acid as required by the experiment. Half of each of the solutions was bottled; the other half was pasteurised by heating to 90°C immediately bottled and cooled. The pair of solutions was chilled (5°C) overnight.

Totally deoxygenated water was prepared by boiling water previously saturated with nitrogen. This was cooled in a nitrogen atmosphere. Deoxygenated sugar solutions were prepared by dissolving sugars in deoxygenated water in a nitrogen atmosphere. As with the aerated solutions, half was kept and half processed (in a nitrogen atmosphere).

(*c*) *Determinations.* Vitamin C was determined by the usual indophenol titration[1] before and after pasteurisation. Sugars were determined by the copper reduction method of Layne and Eynon.[2]

(*d*) *Taste panels.* For fruit juice tasting, panellists were trained, by using various mixtures of processed and unprocessed juice, to differentiate between various degrees of processed flavour. In the tests, panellists were required to rank four samples in order of processed flavour.

With the de-oiled samples, panellists were first asked to distinguish between processed and unprocessed de-oiled orange juice in a triangle test. Panellists were then asked to compare the difference between de-oiled processed and unprocessed orange juice with the difference between processed and unprocessed orange juice.

When tasting sugar solutions, panellists were trained only to distinguish between processed and unprocessed solutions. In the tests, panellists were presented with two unprocessed and one processed solution and asked to nominate the odd sample. The number of panellists nominating the processed solution was taken as a measure of the flavour difference between the processed and unprocessed solutions.

Fruit Juice Results

(*a*) *Fruit juices.* After panellists had ranked four samples of fruit juice from least processed flavour (score 1) to most processed flavour (score 4), it was possible to test for significant agreement between rankings assigned to a number (n) of products by (k) judges. A statistic w, the coefficient of concordance, may be calculated and its significance tested by use of the F distribution:

$$w = \frac{\text{product sum of squares} - (1/k)}{\text{total sum of squares} + (2/k)}$$

and

$$F = \frac{(k - 1)\, w}{1 - w}$$

with $(n - 1) - 2/k$ degrees of freedom for the numerator, and $(k - 1)[(n - 1) - (2/k)]$ degrees of freedom for the denominator.

TABLE 1

Processed Pineapple Juice—Ranking

	Normal ascorbic acid		Elevated ascorbic acid	
	Aerated	Deoxygenated	Aerated	Deoxygenated
Ascorbic acid before processing	3·4 mg/100 g	3·2 mg/100 g	82·7 mg/100 g	82·5 mg/100 g
Total score from ranking	22	16	14	8

$$\text{Correction term} = 60^2/24 = 150$$

$$\text{Product SS} = \frac{(22^2 + 16^2 + 14^2 + 8^2)}{6} - 150 = 17$$

$$\text{Total SS} = 6(1^2 + 2^2 + 3^2 + 4^2) - 150 = 30$$

$$W = \frac{17 - 1/6}{30 + 2/6} = 0{\cdot}56$$

$$F = \frac{(6 - 1)0{\cdot}56}{1 - 0{\cdot}56} = 6{\cdot}4$$

$(4 - 1) - 2/6 = 4{\cdot}67$ Deg of freedom for numerator

$(6 - 1)[4{\cdot}67] = 23{\cdot}35$ Deg of freedom for denominator

F value from Tables[4] assessed as 4·1 at the 1% level

Significant

The calculated values are significant at the 1% level and it may be concluded that the judges exhibit a significant degree of agreement in their ranking of the fruit juices. Also, it can be concluded that both deoxygenation and the addition of extra ascorbic acid leads to less processed flavour on pasteurisation in both pineapple and orange juice.

From Tables 3 and 4, it appears that the amount of ascorbic acid lost in the process is unaffected by the original concentration of ascorbic acid, or the amount of dissolved oxygen in the fruit juice. There is no correlation between the ranking assigned by the panellists and the amount of ascorbic acid lost.

Panellists were presented with a triangle test involving processed and unprocessed de-oiled orange juice and asked to nominate the

TABLE 2

Processed Orange Juice—Ranking

	Normal ascorbic acid		Elevated ascorbic acid	
	Aerated	Deoxygenated	Aerated	Deoxygenated
Ascorbic acid before processing	36·8 mg/100 g	38·3 mg/100 g	105·2 mg/100 g	105·4 mg/100 g
Total score from ranking	21	14	18	7

$$\text{Correction term} = 60^2/24 = 150$$

$$\text{Product SS} = \frac{(21^2 + 14^2 + 18^2 + 7^2)}{6} - 150 = 18$$

$$\text{Total SS} = 6(1^2 + 2^2 + 3^2 + 4^2) - 150 = 30$$

$$W = \frac{18 - 1/6}{30 + 2/6} = 0{\cdot}59$$

$$F = \frac{(6 - 1)0{\cdot}59}{1 - 0{\cdot}59} = 7{\cdot}2$$

4·67 Deg of freedom for numerator

23·35 Deg of freedom for denominator

The value from Tables[4] assessed as 4·1 at the 1% level

Significant

processed sample. They were then asked to compare the difference between processed and unprocessed whole orange juice (given an arbitrary score of 10) with the difference between processed and unprocessed de-oiled orange juice and to attempt to give a score to the difference between the de-oiled samples.

TABLE 3

Ascorbic acid losses and ranking scores on processing pineapple juice

	Ascorbic acid before processing	Ascorbic acid after processing	Loss of ascorbic acid	Total ranking score
Aerated	3·4 mg/100 g	1·9 mg/100 g	1·5 mg/100 g	22
Deoxygenated	3·2 mg/100 g	1·8 mg/100 g	1·4 mg/100 g	16
Aerated + fortified	82·7 mg/100 g	81·0 mg/100 g	1·7 mg/100 g	14
Deoxygenated + fortified	82·5 mg/100 g	80·8 mg/100 g	1·7 mg/100 g	8

TABLE 4

Ascorbic acid losses and ranking scores on processing orange juice

	Ascorbic acid before processing	Ascorbic acid after processing	Loss of ascorbic acid	Total ranking score
Aerated	36·8 mg/100 g	36·6 mg/100 g	0·2 mg/100 g	21
Deoxygenated	38·3 mg/100 g	38·2 mg/100 g	0·1 mg/100 g	14
Aerated + fortified	105·2 mg/100 g	105·0 mg/100 g	0·2 mg/100 g	18
Deoxygenated + fortified	105·4 mg/100 g	105·3 mg/100 g	0·1 mg/100 g	7

TABLE 5

Detection of processed flavour in de-oiled orange juice

	Scores	
(*a*) Triangle test Pick out processed sample—Correct response 100%		
(*b*) Difference between processed/unprocessed de-oiled orange juice [If difference between processed/unprocessed whole orange juice is 10]		
	8	3
	8	7
	5	8
	7	12
Average score = 7·4		

TABLE 6

Detection of processing flavour in model orange juice systems

Solution	% Correct response	Level of significance[4]
Water	30	—
Fructose 2·4 g/100 g	55	0·05
Glucose 2·4 g/100 g	65	0·01
Sucrose 4·8 g/100 g	25	—
Complete sugars as above	75	0·01
Citric acid 1·45 g/100 g	35	—
Complete sugars + citric acid	70	0·01
Ascorbic acid 50 mg/100 g	40	—
Complete sugars + citric acid + ascorbic acid	40	—

As the results in Table 5 show, all the panellists could distinguish between processed and unprocessed de-oiled orange juice. Since the score for the difference between processed and unprocessed de-oiled orange juice was 74% of the score assigned to the same difference for whole orange juice, the results suggest that three quarters of the flavour change which takes place on pasteurisation of orange juice involves the aqueous phase.

Because of the difficulty of removing all the solvent before processing, particularly in a system as complex as orange juice, the figure of 74% cannot be afforded a great deal of significance. However, the result may be taken as an indication that there is at least some flavour change in the aqueous phase and certainly a flavour change in the oil phase on processing.

(*b*) *Model systems.* Since a flavour change appears to take place in the aqueous phase on pasteurisation of fruit juices, it was decided to commence with an entirely aqueous model system. Preliminary results are summarised in Table 6. In each case twenty panellists were employed in triangle tests involving processed and unprocessed solutions. The concentrations employed approximated to orange juice.

Taste panellists proved able to distinguish between processed and unprocessed glucose and fructose solutions and solutions containing glucose and/or fructose along with sucrose and citric acid. When 50 mg of ascorbic acid per 100 g of solution was added,

TABLE 7

% Correct responses for detection of processing flavours in glucose/ascorbic acid solutions

Ascorbic acid mg/100 g	Glucose concentration (g/100 g)				
	0·5	1·25	2·5	3·75	5·0
0	60	75	70	70	65
2	55	60	70	65	60
5	55	60	60	60	55
10	35	40	50	55	55
15	30	40	50	50	50
20	35	35	30	45	45
35	35	35	30	30	30
50	30	30	35	30	40
100	35	35	30	35	35

the flavour difference after processing disappeared. This flavour change was investigated further. The twenty panellists were presented with a series of triangle tests as described. The same glucose concentration was used for a series of tests, the ascorbic acid level being varied. This was repeated for five different glucose concentrations (Table 7).

For a given glucose concentration, the number of correct responses was found to be directly related to the concentration of ascorbic acid (Fig. 1).

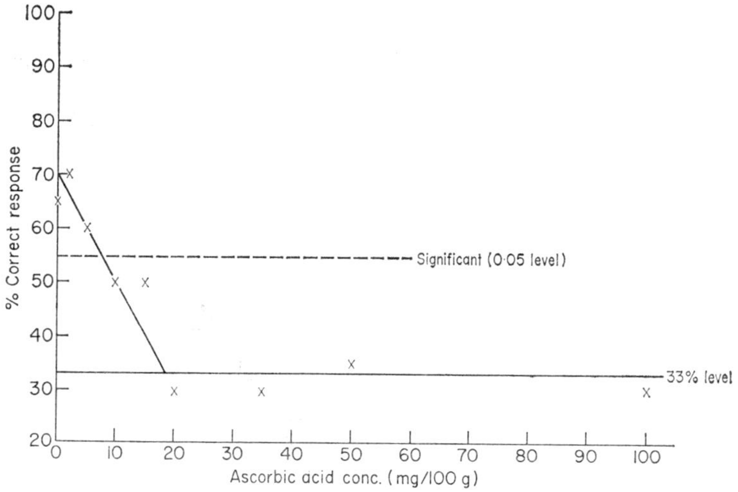

FIG. 1. Percentage correct response versus ascorbic acid concentration in processed ascorbic acid/glucose solutions (2·5 g/100 g glucose solution).

The direct relationship holds until the concentration of ascorbic acid is such that it inhibits any flavour change on pasteurisation and the correct response reaches the random 33% level. When the tests were repeated with different glucose concentrations, a family of straight line graphs emerged (Fig. 2).

From these graphs it is obvious that ascorbic acid is, in some way, acting to prevent the flavour change on pasteurisation. By reference to Fig. 2 it is possible to discover the amount of ascorbic acid which is needed to eliminate the flavour change for a series of glucose concentrations. If the amount of ascorbic acid required is now plotted against glucose concentration, a direct relationship again emerges (Fig. 3).

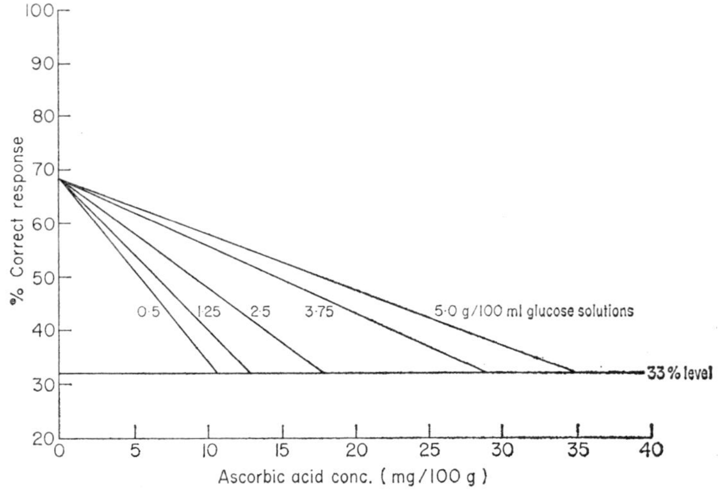

FIG. 2. Percentage correct response versus ascorbic acid concentration for the detection of processed flavour in ascorbic acid/glucose solutions (collated results).

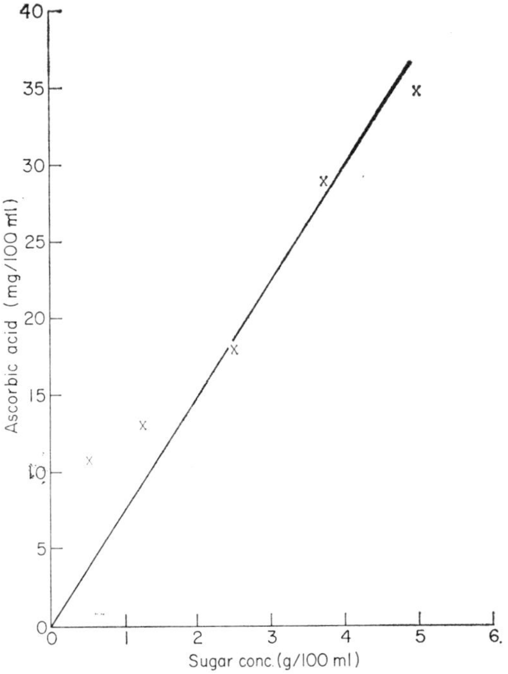

FIG. 3. Ascorbic acid required to eliminate flavour change on processing glucose solutions versus glucose concentration.

No loss of sugar was detectable by the usual volumetric copper reduction method.[2] However, a proportion of the ascorbic acid is lost during the pasteurisation process. Work on this aspect of the pasteurisation process is embryonic, but indications are that the amount of ascorbic acid lost on processing is related both to the original concentration of ascorbic acid and to the glucose concentration.

When panellists were presented with triangle tests involving deoxygenated processed and unprocessed glucose solutions, they proved unable to detect a difference. Significant numbers of panellists (0·05 level) could distinguish between aerated processed and deoxygenated processed glucose solutions. Oxygen is thus implicated in the flavour change.

Discussion of Fruit Juice Results

The mechanism of the flavour change has not been elucidated; possibly very small amounts of glucose (or fructose) are converted to a relatively highly flavoured compound. It has in any case been shown that oxygen is necessary for the flavour changing reaction. The solubility of oxygen decreases sharply as the temperature approaches 90°C and it is probable that the amount of oxygen available at high temperatures is the limiting factor in flavour production.

When ascorbic acid is introduced into the system, the well known ascorbic acid to dehydro-ascorbic acid reaction can occur in the presence of oxygen (Fig. 4).[5–8]

The ascorbic acid will thus act as an oxygen scavenger, removing the small amounts of oxygen present, preventing the oxidation of the reducing sugars and hence preventing the flavour change.

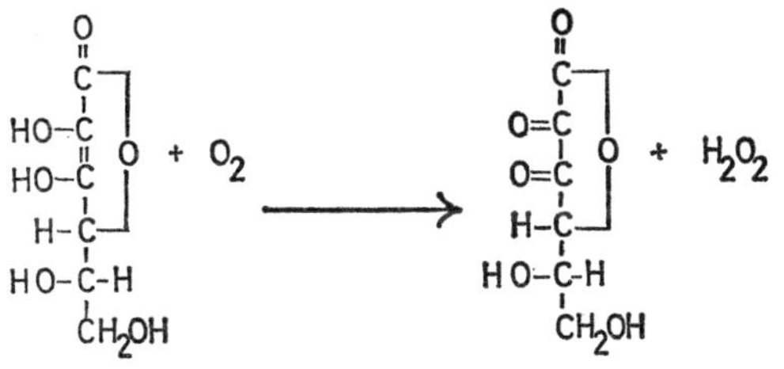

L-ascorbic acid. Dehydro-ascorbic acid.

FIG. 4.

Flavour changes of the type described could be of importance in the pasteurisation of fruit juices. From a knowledge of the reducing sugar content of a given juice, it is possible to predict the amount of ascorbic acid required to prevent this flavour change (*see* Fig. 3).

Examination of the reducing sugar and ascorbic acid contents of the fruit juices commonly produced in this country show that this type of flavour change should only occur in pineapple, apple and tomato juice to which extra ascorbic acid has not been added, and other juices exceptionally low in this vitamin (Table 8).

The fortification of apple juice with ascorbic acid has already been recommended to prevent browning of the juice on processing.[11,12]

Conclusions from Fruit Juice Results

(1) The addition of ascorbic acid and deoxygenation prior to pasteurisation help to reduce 'processed flavour' in the production of both pineapple and orange juice.
(2) Glucose and fructose solutions undergo a change in flavour on heating which does not occur with sucrose solutions.
(3) Oxygen is essential to this flavour change.
(4) The flavour change in glucose solutions can be reduced or eliminated by the addition of ascorbic acid.
(5) The amount of ascorbic acid required to eliminate the flavour change is proportional to the glucose concentration.
(6) The applicability of these model studies to fruit juice processing is dependent on type. Grapefruit, lemon and orange appear to contain sufficient ascorbic acid to exert the necessary protective effect. Apple, pineapple and tomato juices may require ascorbic acid addition (*see* Table 8).

TABLE 8

Juice	Reducing sugars[9] (%)	Ascorbic acid required (Fig. 3) (mg/100 g)	Ascorbic acid[10] present (mg/100 g)
Apple	8·3	64·0	5
Grapefruit	3·2	23·5	35
Lemon	1·4	10·5	50
Orange	5·1	38·5	50
Pineapple	4·2	32·0	8
Tomato	3·4	26·0	16

QUALITY CHANGES DURING PEA PROCESSING

The main factors that may affect the loss of vitamin C during blanching may be summarised as follows:

(A) **Factors Affecting the Initial Vitamin C Content of Peas at Blanching**

(1) Variety of peas.
(2) Growing conditions.
(3) Botanical maturity of the peas at the optimum harvest time.
(4) Damage to peas caused by vining method.
(5) Delay time and ambient temperature between vining and the pre-blanch operations.
(6) Damage and leaching caused by such pre-blanch operations as cleaning and washing.

(B) **Factors Affecting the Loss of Vitamin C During Blanching**

(1) Size of peas.
(2) Product to water ratio.
(3) Amount of stirring of blanch water.
(4) Blanch temperature.
(5) Blanch time.
(6) Soluble solids content of the blanch water.
(7) Amount of post-blanch water cooling.

A review of previous work on pea-blanching revealed that many of these factors had been ignored.[13] Therefore, in order to facilitate control over the above variables, peas were specially grown and blanching operations were carried out on a small scale in order to achieve accurate observation of the location and state of the vitamin C as a result of blanching.

Now not only is vitamin C easily extractable because of its high water solubility, but it is also susceptible to both chemical and enzymic oxidation. Ascorbic acid may be oxidised to dehydro-ascorbic acid which is more heat labile than ascorbic acid and is also easily hydrolysed to 2,3-diketogulonic acid at a rate dependent on pH and temperature.[14] The relative amounts of ascorbic and dehydroascorbic acid prior to heat processing are important as dehydroascorbic acid is nutritionally as active as ascorbic acid, whereas

2,3-diketogulonic acid has no vitamin C activity.[15] Most previous work on this subject has been concerned only with ascorbic acid, possibly because the dehydro-ascorbic acid content of fresh peas is small. However, the action of enzymes and heat during handling and processing could give rise to much larger quantities of dehydro-ascorbic acid and therefore an even greater loss of vitamin C.

Olson[16] has observed the relative amounts of ascorbic, dehydro-ascorbic and diketogulonic acids in strawberries stored at −7°C for 120 days and found that while the sum of the three compounds remained virtually constant, the diketogulonic acid content increased linearly with time. Therefore it was thought feasible that a relationship between these compounds might exist in peas, which would vary according to the various processes undergone.

Experimental Methods

(*a*) *Growing and sampling.* Randomised plots of Dark Skinned Perfection peas were grown in a sandy loam near Chertsey in Surrey. Pods were removed by hand from the first two nodes of randomly selected vines at the optimum harvest time for freezing, which was determined by a combination of factors including accumulated heat units and the alcohol insoluble solids content. The peas were then shelled in the laboratory, and no detectable change in the vitamin C content was observed during the period prior to blanching. Peas with a diameter between 8·7 and 10·3 mm were selected for blanching, the equivalent tenderometer reading being in the range 95–105. By using such a sampling procedure it was hoped to minimise the vitamin C content variation between samples so that results could be directly compared. The actual variation is shown in Table 9.

(*b*) *Blanching procedure.* In order to simulate a commercial product to water ratio of one to four, 40 peas (20 g) were blanched in 80 ml boiled distilled water in a 250 ml beaker, the water being gently agitated by a magnetic stirrer rotating at 78 rpm. It was assumed that continuous commercial water blanchers steadily maintain the required blanch temperature; therefore in order to minimise the temperature drop caused by addition of the peas to the water, after pea addition, the beaker was placed over a bunsen flame for 10 s before returning to an electrically heated hot plate set to maintain the required blanch temperature to within 0·5°C.

After the required time, the peas were sieved out over a funnel and the blanch water collected in an ice cooled 100 ml volumetric flask,

TABLE 9

Variation of vitamin C content between samples of 40 peas

Sample number	Vitamin C content mg/100 g					
	Wet wt. basis			Dry wt. basis		
	Ascorbic acid	Total vit. C	Dehydro-ascorbic acid (by difference)	Ascorbic acid	Total vit. C	Dehydro-ascorbic acid (by difference)
1	32·5	35·8	3·3	154·7	170·4	15·7
2	30·9	32·4	1·5	147·1	154·2	7·1
3	31·9	34·1	2·2	151·8	162·3	10·5
4	30·5	32·9	2·4	145·2	156·6	11·4
5	30·5	33·5	3·0	145·2	159·5	14·3
6	31·5	33·3	1·8	149·9	158·5	8·6
7	30·5	33·0	2·5	145·2	157·1	11·9
8	30·3	31·0	0·7	144·2	147·6	3·4
9	30·7	34·5	3·8	146·1	164·2	18·1
10	32·5	35·2	2·7	154·7	167·6	12·9
Mean	31·2	33·6	2·4	148·4	159·8	11·4
Standard deviation	0·7	0·5	—	4·2	6·3	—
Standard error of mean	0·2	0·2	—	1·3	2·0	—
Coefficient of variation	2·2	1·5	—	2·8	3·9	—

the solution being made up to the mark with 6% metaphosphoric-acetic acid. The peas were immediately cooled in 50 ml ice cold 6% metaphosphoric-acetic acid[17] ready for maceration. It should be noted that the post-blanching cooling procedure was regarded as a separate stage in these studies and as yet has not been included.

(*c*) *Determination of vitamin C*. The peas were macerated with the 50 ml extractant in a Waring Blendor for one minute before centrifuging at 2650 rpm for 4 min in 40 ml centrifuge tubes. Centrifuging was repeated after resuspending in further extractant, and the supernatant was filtered through muslin and made up to 200 ml with extractant. This rapid procedure rendered a 95% extraction of vitamin C. The ascorbic acid was then determined by titration with 2,6-dichlorophenolindophenol dye and the total vitamin C (ascorbic acid plus dehydro-ascorbic acid) by the fluorimetric method as described by the AOAC.[17]

Cotyledons and testa were separated with the aid of a scalpel dipped in metaphosphoric-acetic acid, and extraction carried out as before, but finally making up to 100 ml.

(*d*) *Determination of alcohol insoluble solids*. This was carried out on 20 g samples by the method of Moyer and Holgate,[18] but drying was done at 103–105°C in order to complete the determination on the same day.

(*e*) *Determination of moisture content*. This was achieved by drying 20 g pea samples to constant weight at 103–105°C.

(*f*) *Enzyme extraction procedure*. 20 g peas were macerated with 50 ml ice cold 0·4 molar sodium chloride solution[19] for 1 min in a Waring Blendor and left for 2 hr at 1°C with occasional shaking. The macerate was then filtered through muslin to facilitate pipetting at a later stage, and made up to 200 ml with 0·4 molar sodium chloride. For measurement of peroxidase activity, the solution was filtered through Whatman 42 paper.

(*g*) *Determination of peroxidase activity*. The method of Pinsent[20] was adapted for use on a Unicam SP800 Spectrophotometer. The residual activity after blanching was measured, to indicate the heat treatment given by the blanching procedure adopted (Table 10).

(*h*) *Determination of ascorbic acid oxidase activity*. This was measured by reacting a known volume of the enzyme extract with a known volume of a standard ascorbic acid solution for a fixed time, and estimating the ascorbic acid remaining after that time. All the solutions were warmed to 30°C prior to the start of the reaction, and

TABLE 10

Residual peroxidase activity in whole peas after blanching at the given temperature

97°C		85°C		70°C	
Blanch time (s)	Peroxidase activity (%) in fresh peas	Blanch time (s)	Peroxidase activity (%) in fresh peas	Blanch time (min)	Peroxidase activity (%) in fresh peas
0	100	0	100	0	100
15	41	15	47	2	72
30	5	45	18	5	64
45	0	90	0	10	52
—	—	—	—	15	35
—	—	—	—	20	36
—	—	—	—	25	27

were all made up with glass distilled water. 35 ml of a phosphate–citrate buffer at pH 5·9, the pH of the pea macerate, were placed in a 100 ml conical flask in a 30°C shaker bath. 5 ml of an ascorbic acid solution containing 300 μg of ascorbic acid per ml were added to the flask. The reaction was then started by adding 10 ml of the enzyme extract. After 5 min the reaction was stopped by the addition of 20 ml ice cold 6% metaphosphoric-acetic acid. The contents of the flask were then made up to 100 ml with more acid and aliquots titrated against 0·025% indophenol dye. Under the given reaction conditions, not more than 20% of the initial substrate was oxidised.

Pea Blanching Results

Location of vitamin C in the pea. It has been observed that ascorbic acid is most highly concentrated in the more actively metabolising regions of plant tissue,[15] and it is known that the vitamin C content of the pea decreases with maturity and that the cotyledon to testa (skin) ratio increases with maturity.[21] Also Reeve[22] has observed that tissue resorption occurs along the inner surface of the testa, and that a highly specialised epidermal structure composed of macrosclereid cells is formed and pronounced wall thickenings of a pentosan-cellulosic composition are developed causing toughening

TABLE 11
Location of vitamin C in peas

Location	Alcohol insoluble solids (%)	Equivalent tenderometer reading	Average % of fresh weight	Average dry weight (%)	Vitamin C (mg/100 g)			
					Wet wt. basis		Dry wt. basis	
					Ascorbic acid	Total vit. C	Ascorbic acid	Total vit. C
Whole peas	11·13	105	100·0	21·3	30·6	32·9	143·7	154·5
Cotyledons	—	—	70·1	20·7	22·4	24·2	108·2	116·9
Testa	—	—	29·9	22·1	49·9	53·2	225·8	240·7

during maturation. Therefore it was thought probable that the testa would contain a relatively high vitamin C content. This was found to be so, as indicated in Table 11, the testa tissue containing approximately twice as much vitamin C as the cotyledonous tissue, weight for weight, at this maturity.

(*a*) *Loss of ascorbic acid and soluble solids into the water*. Blanching was carried out at 97, 85 and 70°C, and the ascorbic acid and total vitamin C (ascorbic plus dehydro-ascorbic acid) remaining in the peas and the blanch water were determined. The increase in the ascorbic acid content of the blanch water is shown in Fig. 5. No significant loss of ascorbic acid due to chemical oxidation was observed in the water at the temperatures and times used. The increase of soluble solids in the water was also measured, as shown in Fig. 6.

(*b*) *Location and loss of ascorbic acid by oxidation*. Since there was a higher vitamin C concentration in the testa, further tests were carried out in which the cotyledons and testa were separated after blanching at 97 and 70°C, and the ascorbic acid and total vitamin C measured in the separated tissues. The location of the ascorbic acid is expressed as a percentage of the initial ascorbic acid content, and an average of two blanches is shown for clarity in Figs. 7 and 8.

(*c*) *Activity of ascorbic acid oxidase*. An indication of the relative activity of the enzyme at different temperatures was obtained using the standard 5 min reaction time as previously described (Fig. 9).

In order to obtain a comparison of the activity of the extracted enzyme and the activity of the enzyme in the intact pea, peas were blanched for 10 min at different temperatures, and the ascorbic acid measured in the peas and blanch water, the amount of ascorbic acid oxidised being calculated by difference from the initial ascorbic acid content of the peas, as shown in Fig. 10.

Discussion of Pea Blanching Results

(*a*) *Loss of ascorbic acid and soluble solids into the water*. Considerable amounts of ascorbic acid were found in the water, but no detectable amounts of dehydro-ascorbic acid were observed. A 1 min blanch at 97°C resulted in a loss of 28 % of the initial ascorbic acid in the pea (*see* Fig. 7). It would be expected that if the heat treatment was sufficient to inactivate the ascorbic acid oxidase, any ascorbic acid loss would be accounted for in the blanch water. And it was found that when blanching at 97 and 85°C, the sum of the

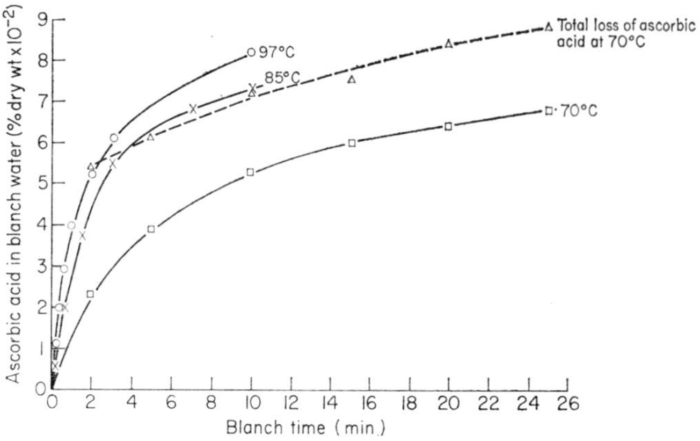

FIG. 5. Ascorbic acid lost into blanch water at the given temperature.

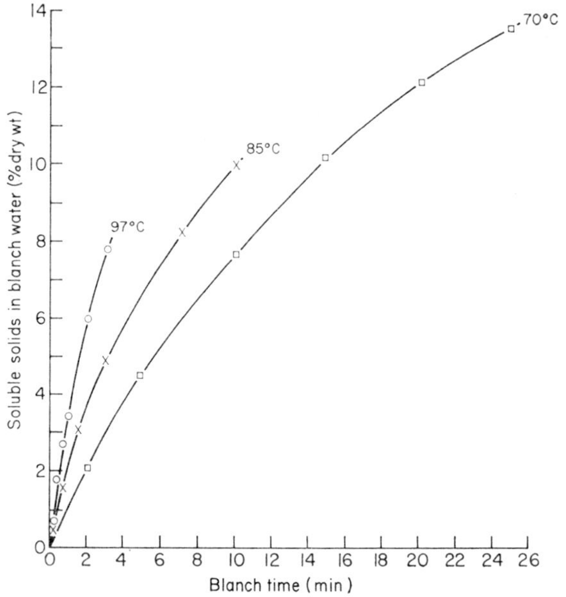

FIG. 6. Soluble solids lost into the blanch water at the given temperature.

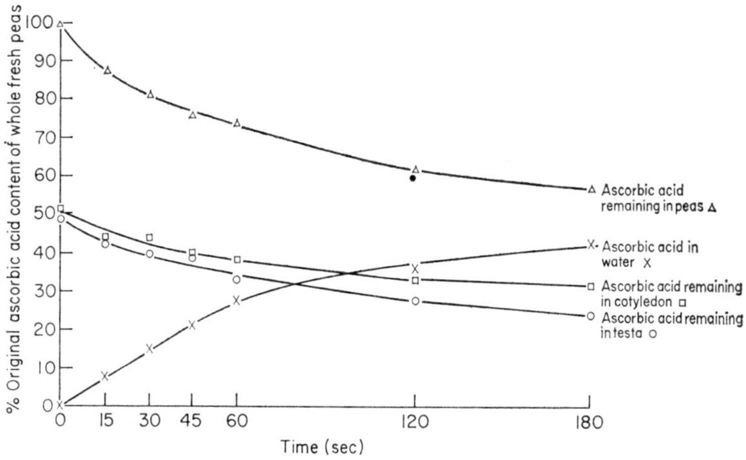

FIG. 7. Percentage of the original ascorbic acid content of whole fresh peas remaining in the whole peas, the cotyledons, the testa, and in the blanch water after blanching in distilled water at 97°C.

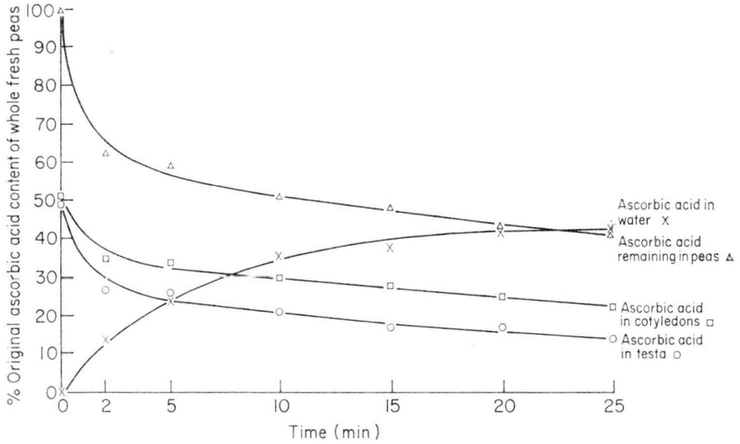

FIG. 8. Percentage of the original ascorbic acid content of whole fresh peas remaining in the whole peas, the cotyledons, the testa, and in the blanch water after blanching in distilled water at 70°C.

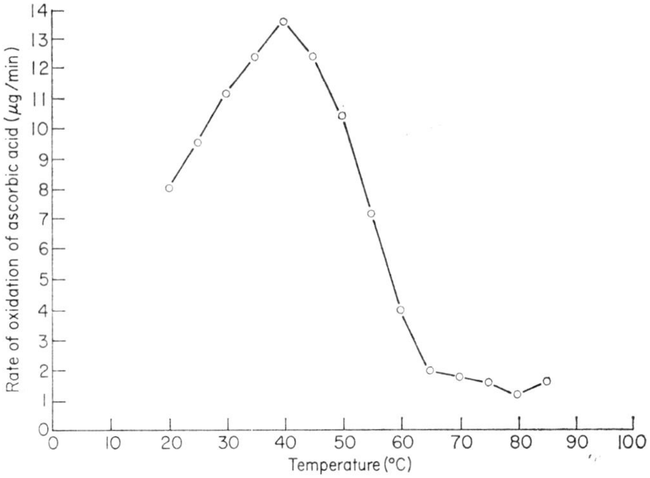

FIG. 9. Variation of ascorbic acid oxidase activity with temperature.

ascorbic acid remaining in the peas and the ascorbic acid in the water virtually equalled the initial ascorbic acid content of the peas. But at 70°C, the total loss was not accounted for solely in the water (*see* Fig. 5). This shows that at the two higher temperatures virtually no oxidation had occurred which could be detected outside experimental error, and that some 24% of the initial ascorbic acid present had been oxidised in the first 2 min at 70°C (*see* Figs. 5 and 8).

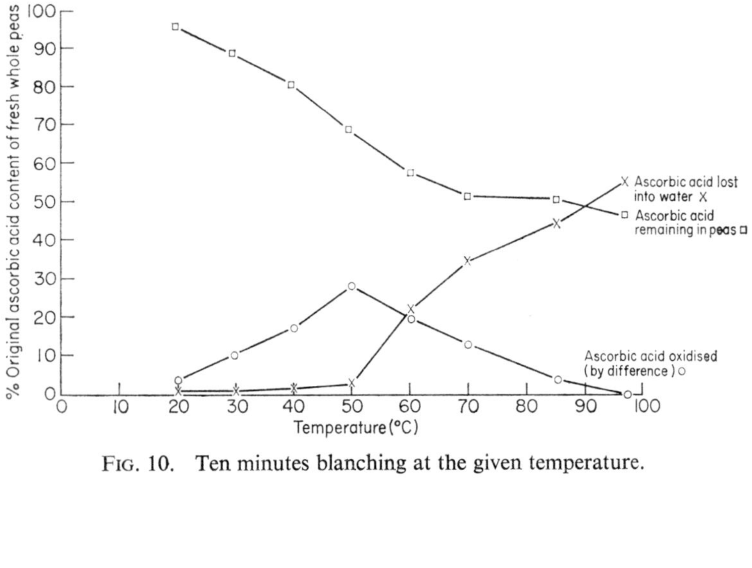

FIG. 10. Ten minutes blanching at the given temperature.

Because of this oxidative loss at 70°C, the total loss of ascorbic acid after 3 min is similar at all three temperatures.

As the vitamin C concentration is higher in the testa, approximately 51 % of the ascorbic acid content of the whole pea is located in the cotyledons and 49 % is located in the thin outer testa. Therefore it would be expected that the initial rate of loss of ascorbic acid would be significantly higher than the final rate, and that the rate of loss of soluble solids would decrease more steadily with time. The curves obtained in Figs. 5 and 6 suggest that this is so, a significant change in the ascorbic acid loss rate occurring when 30–35 % of the ascorbic acid has been lost. And taking the 85°C blanch as an example, it is found that the rate of loss of ascorbic acid after 1 min blanching is 16·6 times higher than the rate after 10 min blanching; whereas the rate of loss of soluble solids is only 3·7 times higher.

(*b*) *Location and loss of ascorbic acid by oxidation.* It has already been seen that the oxidation occurring at 97 and 85°C was negligible, but was considerable at 70°C. Now the thermal half-life of dehydro-ascorbic acid at pH 6 is less than 1 min at 100°C, and is 2 min at 70°C, irrespective of oxygen concentration. At the common time of 2 min, it would be expected that no detectable amount of dehydro-ascorbic acid would be present at 97°C because of the small amount of dehydro-ascorbic acid present initially, but that at 70°C, dehydro-ascorbic acid would be present in an amount equal to the total vitamin C destroyed, as estimated by difference. This assumes that the oxidation taking place during the 70°C blanch occurs within the first few seconds of blanching, *i.e.* before the ascorbic acid oxidase is inactivated while the temperature of the pea is rising to 65°C (*see* Fig. 9). Unfortunately, the problems of measuring the pea centre temperature were not overcome during these studies.

The total vitamin C figures after 2 min blanching are shown in Table 12. At 70°C 11 % of the initial total vitamin C content has been destroyed, as 74 % remains in the peas and 15 % in the blanch water. And as 62 % of the total vitamin C remaining is ascorbic acid, by difference, 12 % of the total vitamin C remaining is dehydro-ascorbic acid. This shows that within the first few seconds of blanching, 23 % of the total vitamin C was present as dehydro-ascorbic acid, and approximately 16 % was formed as a result of enzyme activity during that time, since the initial dehydro-ascorbic acid content of the peas was 7 %. A similar pattern is observed in the separated cotyledons and testa, although the experimental error caused by separation may be greater than the 2% error of the other values.

It is also seen that the loss of ascorbic acid is rapid in the first 2 min (*see* Fig. 8) and that the percentage loss from the testa tissue is greater, as would be expected.

TABLE 12

Comparison of location of vitamin C after 2 min blanching at 97°C and 70°C. (Both ascorbic acid and total vitamin C are expressed as % of the initial total *vitamin C content of the fresh whole pea)*

	Whole peas				Blanch water			
	Ascorbic acid		Total vitamin C		Ascorbic acid		Total vitamin C	
Temp. °C	97	70	97	70	97	70	97	70
Vit. C remaining (%)	60	62	63	74	33	15	33	15
	Cotyledons				**Testa**			
	Ascorbic acid		Total vitamin C		Ascorbic acid		Total vitamin C	
Temp. °C	97	70	97	70	97	70	97	70
Vit. C remaining (%)	31	34	33	44	26	27	30	35

(*c*) *Activity of ascorbic acid oxidase.* Figure 9 shows that maximum activity occurs at about 40°C, and at 60°C the activity is half that at 20°C. Above 65°C the activity is negligible. This emphasises the requirement that to minimise vitamin C oxidation the rate of heat input to the pea must be fast, so that the pea centre temperature reaches 65°C as rapidly as possible. From this curve, it would be expected that the amount of ascorbic acid oxidised in whole peas, when blanched at the same temperatures, would follow a similar trend. However, in the organised structure of intact pea tissue, the enzyme and substrate would not be as freely in contact until the cell structure broke down due to the action of heat. Therefore the fastest rate of oxidation should occur at the point of maximum cell disruption with minimum enzyme inactivation.

Figure 10 shows that this optimum point occurs at about 50°C. Here the leaching losses increase rapidly as the cells die and the cell membranes become permeable. Figure 10 also suggests that a significant change in the mode of vitamin C loss during blanching might occur if the peas were cut or bruised prior to blanching, which is the case in commercial operations as a result of the mechanical shelling of peas.

Holdsworth[23] has drawn attention to the problems of handling and vining peas, and it is clear that the combination of bruised peas and a high ambient temperature prior to blanching, could give rise to losses during blanching that are much greater than those which have been observed in these studies. Further work has already been started on the effects of damage on the loss of vitamin C from peas on processing.

Conclusions from Pea Blanching Results

(1) Leaching is the prime cause of vitamin C loss when undamaged peas are blanched at 97°C, but significant losses due to oxidation occur at 70°C.
(2) There is no useful relationship between the relative amounts of ascorbic, dehydro-ascorbic and diketogulonic acids in peas and the blanching time and temperature.
(3) Vitamin C losses due to enzymic action, both prior to and during blanching, may be very significant if the pea tissue is damaged. Further work is required to determine the effects of damage to peas on vitamin C losses during blanching.
(4) Further work is needed to evaluate the other factors affecting the pea blanching operation using the standard procedures adopted, in order to obtain the optimum blanching conditions.
(5) These studies do not fully indicate the mechanisms of loss due to leaching, and further experimental work is required. It is suggested that for such work, a more uniform plant tissue should be used, such as is found in carrots.

CONCLUSIONS

The unique oxidation–reduction system of L-ascorbic acid and its metabolites clearly has a fundamentally significant role in modern fruit and vegetable technology. The work described in this twofold project underlines the immensity of detailed analyses which must be accumulated before the diverse effects of the ascorbate system on food quality can be fully evaluated.

ACKNOWLEDGEMENTS

The authors are grateful to D. A. Donnett (Genete) Ltd for providing the fruit juices, and to Birds Eye Foods Ltd for providing the pea seed. Mr H. W. Symons (Birds Eye) and Miss J. Salmon (Birds Eye) are also thanked for their help and advice.

REFERENCES

1. Pearson, D. (1970). *The Chemical Analysis of Foods,* J. A. Churchill, London.
2. Layne, J. H. and Eynon, L. (1923). *J. Soc. Chem. Ind. Lond.,* **42,** p. 32T.
3. Kendal, M. G. (1948). *Rank Correlation Methods,* Charles Griffin and Co., London.
4. Amerine, M. A., Pangborn, R. M. and Roessler, E. B. (1965). *Principles of Sensory Evaluation of Food,* Academic Press, New York.
5. Stroehecker, R. and Schmidt, H. (1943). *Z. Lebensm. Untersuch. Forsch.,* **86,** p, 370.
6. Scaife, J. F. (1959). *Can. J. Biochem. Physiol.,* **37,** p. 1049.
7. Butt, V. S. and Holloway, M. (1961). *Arch. Biochem. Biophys.,* **92,** p. 24.
8. Dawson, C. R. (1960). *Ann. N.Y. Acad. Sci.,* **88,** p. 353.
9. Harding, M. G. *et al.* (1965). *J. Amer. Diet. Ass.,* **46,** p. 197.
10. McCance, R. A. and Widdowson, E. M. (1960). *The Composition of Food,* H.M. Stationery Office, London.
11. Tressler, D. K. and Joslyn, M. A. (1961). *Fruit and Vegetable Processing Technology,* Avi Publishing Co., Westport, Connect. USA.
12. Cruess, W. V. (1958). *Commercial Fruit and Vegetable Products,* McGraw-Hill, New York.
13. Lee, F. A. (1958). *Adv. Food Res.,* **8,** p. 63.
14. Huelin, F. E. (1949). *Austr. J. Sci. Res. Series B Biol. Sci.,* **2**(4), p. 346.
15. Sebrell, W. H. and Harris, R. S. (Eds.) (1967). *The Vitamins,* 2nd Edn, Vol. 1, Academic Press, New York and London.
16. Olson, R. L. (1960). *Conference on Frozen Food Quality,* p. 30. Held at Western Reg. Res. Lab. Albany, California on 4–5 Nov., Agric. Res. Service 74–21 USDA.
17. *Official Methods of Analysis* (1970). 11th Edn, Assoc. of Offic. Anal. Chem., Washington, D.C. 20044.

18. Moyer, J. C. and Holgate, K. C. (1948). *Anal. Chem.*, **20**(5), p. 472.
19. Huelin, F. E. and Stephens, I. M. (1948). *Austr. J. Sci. Res. Series B Biol. Sci.*, **1**(1), p. 58.
20. Pinsent, B. R. W. (1962). *J. Food Sci.*, **27**, p. 120.
21. McKee, H. S., Robertson, R. N. and Lee, J. B. (1955). *Austr. J. Biol. Sci.*, **8**, p. 137.
22. Reeve, R. M. (1949). *Food Res.*, **14**, p. 77.
23. Holdsworth, S. D. (1969). *Proc. Biochem.*, **4**(7), p. 26.

DISCUSSION

Chairman: I was interested in your comments on pasteurisation of fruit juice because there is one system in the world at least which uses immediately after pasteurisation a chilling system, and then a concentration process carried out in a Linde–Krauser drum, or Krauser–Linde drum. In fact the retention of vitamin C is very high and certainly the juice has a very good flavour. I don't know if you had time to look into that.

Bointon: I found that losses of vitamin C were very low even using a pasteurisation process which involved the deliberate aeration of the juice beforehand. But flavour changes were certainly more pronounced when oxygen was present.

Broomfield: On the question of the flavours of fruit juices, could you tell us what period of time had elapsed between heat treatment and the tasting of the samples?

Bointon: The samples were pasteurised in the evening, cooled and refrigerated overnight and tasted the following morning.

Broomfield: Were the samples tasted again after a given length of storage—say a week or a fortnight—and if so were there any differences in the assessments when the results were compared with those obtained when freshly pasteurised juices were tested against fresh juices?

Bointon: The short answer is no. I was trying to separate actual pasteurisation effects from storage effects. I think there has been much more work done on storage effects and so I was not working on that at all.

Ford: When you use the term 'off-flavour' I think you imply that your panellists disliked the flavour. Certainly the appreciation of flavour in milk, sterile milk, is a very subjective thing—for instance, the cooked flavour of sterilised milk is very much liked in the Midlands.

Bointon: I think 'off-flavour' is probably a bad term to use. In fact, the first experience I had of taste testing was between pasteurised juice and unpasteurised juice. Panellists were asked to pick out the best flavours, and they all went for the pasteurised juice, because this is what people are used to. But I think you will agree there is technically a flavour change.

Questioner: Have you noticed any change of nutritional significance?

Bointon: I haven't really investigated the nutritional change. I think the main nutritional use of fruit juices is for their vitamin C content and sc I would think that is the only possible change.

Wilson: Do you think that the effect of the vitamin C is due to an interaction of the vitamin C with the fruit or whatever it is in the tin, or do you think it is due to an effect of the vitamin C on the taste receptors in the tongue?
Bointon: I think it's unlikely to be a direct effect on the taste buds, because the levels of ascorbic acid used were below threshold levels.
Wilson: Did you measure the plasma ascorbic acid levels in your subject?
Bointon: No.
Pollitt: What is a triangle test, and what is a tenderometer?
Bointon: A triangle test is a fairly standard taste trial procedure, where you present panellists with three samples, two of which are the same and one different, and they are merely asked to nominate the odd sample.
Selman: The tenderometer is a measure of maturity of peas that is used commercially—measuring the hardness value, if you like. It literally means you take a sample of peas and squash them, measuring the force it takes.
Cooke: Does the addition of vitamin C to the juices mentioned sometimes induce discoloration owing to reaction with the juice colour components?
Bointon: I didn't in fact measure the colour. We didn't really find that the amount of vitamin C we added to the juice made an awful lot of difference.
Questioner: With or without aeration?
Bointon: The ones with aeration certainly had a slightly worse colour, but I don't think we could distinguish between the two aerated samples that were fortified and not fortified.
Chairman: I think we've got to take into account the short period of time involved. I'm sure if it had been a longer period, it would have been more noticeable.
Alexander: Could you tell me the sources of the orange and other juices used in the test, and what were the pasteurisation conditions?
Bointon: The orange juice was frozen concentrate from Israel. I used a deliberately harsh pasteurisation—in most cases I aerated before I pasteurised, to make sure the oxygen was present, and I pasteurised by heating to 90°C and then cooling immediately afterwards. I didn't investigate the rate at which the juices cooled down.
Chairman: I think in fact the source you used there, Israeli concentrate, is probably the lowest in vitamin C of any of the orange juices.

GENERAL DISCUSSION

Lewin: I did understand you to say that the ascorbic acid was hydrolytically broken down to diketogulonic acid. Now you can break down ascorbic acid hydrolytically, but the addition of 8% HCl can be used to re-make the ring structure. How did you distinguish between a ring-structured vitamin C and a hydrolytically ruptured ring structure?
Selman: I only measured the ascorbic acid and dehydro-ascorbic acid as such, I didn't measure the diketogulonic acid.

Lewin: You don't break up ascorbic acid hydrolytically to diketogulonic acid. You break it up hydrolytically and the next stage is oxidation to diketogulonic acid. If you avoid the oxidation to diketogulonic acid, you have an acyclic structure. How do you allow for this? It is not biologically active, because to be biologically active it requires the ring structure of ascorbic acid. This problem has been running throughout the symposium this morning and no one has tackled it.

Everybody talks about ascorbic acid nicely and conveniently. The trouble is the molecular biologist looks at it and says what are you doing? What is it? Ascorbic acid has a ring structure. If you destroy the ring structure it is no longer biologically active. Here you have hydrolytically destroyed the ring structure but you have an oxidised product on the right-hand side. My question to you is, how do you distinguish between the ascorbic acid system on the left-hand side, which contains the ring structure, and the open ring structure on the right-hand side (*see* Fig. 1, page 224) to which literature conveniently doesn't give any name but to which if you add 8% HCl (and heat it at 50°C) you will get back your ascorbic acid? Now this is a very serious question. I am asking you—how do you do it? There is a source of error here which unfortunately people have ignored, and which has been triggered by the statement this morning that you hydrolytically break it down into diketogulonic acid, which you don't do: you break it up into the compound on the right-hand side, and you then oxidise it to diketogulonic acid.

I think this is a very serious problem, which shouldn't be avoided, because some of the sources of error in vitamin C assay are due to this.

Long: Could you tell me what sort of concentrations of vitamin C were added to the fruit juices over and above the amount present naturally or after processing?

Bointon: The levels of ascorbic acid in the pineapple juice were very low, something like 5 mg/100 g. The Israeli orange concentrate was also very low—when reconstituted it was about 30–35 mg/100 g. Then I fortified these by adding about 50 mg to each, so that the final juices contained about 55 and about 85 mg respectively.

Coultate: What is the nutritional purpose of adding vitamin C to dried milk powder?

Kläui: There is certainly no question about absorption. The reason for adding it in general is more on the technological side—as a stabilising agent.

5

Nutritional Interactions Between Vitamin C and Heavy Metals

R. E. HUGHES

University of Wales Institute of Science and Technology, King Edward VII Avenue, Cardiff, Wales

ABSTRACT

Inter-relationships between ascorbic acid and metals may be either direct or indirect (mediated, e.g. *via thiol groupings) and may influence either gastro-intestinal absorption or distribution pattern in the tissues, or both. In the case of iron, there is evidence that ascorbic acid can influence both the absorption of iron and its mobilisation and retention by the tissues.*

The pattern for other metals is less clearly defined. This paper will describe the effect of ascorbic acid upon the absorption of cadmium and mercury, the effect of metals upon tissue ascorbic acid, possible relationships between thiol groups, ascorbic acid and tissue levels of potentially toxic metals, and the significance of these relationships vis-à-vis *high and low intakes of ascorbic acid.*

There is some evidence that trace elements such as manganese, cobalt and molybdenum may influence the tissue levels of ascorbic acid (L-xylo-ascorbic acid, vitamin C) in rats and in other species not dependent upon dietary sources of the vitamin.[1,2] Relationships of this type are primarily attributable to modifications induced in the ascorbic acid biosynthetic pathway and as such carry no relevance for man and other species unable to produce their own ascorbic acid. The following discussion deals with relationships between certain metals and pre-formed ascorbic acid and derives primarily from studies on man and the guinea-pig. A true evaluation of these relationships is made difficult by a number of factors. There is evidence not only that metals may influence the distribution and concentration of ascorbic acid but also that ascorbic acid itself may influence the physiological fate of metals by a quite separate type of mechanism. One result of this is that to date studies in the field have barely advanced beyond the state of a series of conceptually unrelated *ad hoc* experiments. Again, much published work has been based on the behaviour of metals in grossly scorbutic animals; during the final stages of scurvy in guinea-pigs inanition and tissue breakdown render invalid any interpretation which seeks to relate metabolic changes to simple ascorbic acid deficiency. Hypovitaminotic animals, rather than scorbutic ones, should always be used in studies of this type.

There are, in physiological terms, two main areas where interplay between metals and ascorbic acid is likely to be of significance. The first area is membrane transport, and more specifically, gastro-intestinal absorption. The second is the retention, mobilisation and metabolism of one or both groups of participants at the tissue level. The bulk of published work relates to the relationships between ascorbic acid and iron although recent studies have indicated that the relationship between ascorbic acid and some non-nutrient metals (such as mercury) could also be of some significance.

ABSORPTION OF METALS

It is fairly generally accepted that ascorbic acid enhances the absorption of iron.[3-5] Despite a considerable output of papers there is still no general agreement on the mechanism of iron absorption.[6-9] There is evidence that the absorption is enhanced by reducing agents such as hydroquinone, cysteine and ascorbic acid.[7,8] The earlier supposition that reducing agents facilitated absorption by maintaining the iron in the reduced (ferrous) state has recently been questioned. Thus both hydroquinone and cysteine enhance the absorption of iron even when sufficient ascorbic acid is present to maintain the iron permanently in the ferrous state.[10] It was consequently suggested that reducing agents aided the absorption of iron, not by maintaining it in the reduced state but by shifting the redox balance within the absorptive epithelium in the direction of reduction, which in some unexplained manner results in an increased absorption.[11] Other workers have drawn attention to the possibility that ascorbic acid functions by virtue of its ability to act as a chelating agent.[9,12,13]

More recently, interest has centred on the possible existence of a carrier or acceptor molecule in the transfer of iron from the intestinal epithelial cell to the plasma. Ferritin, transferrin and a hitherto incompletely characterised carrier molecule have all in turn been suggested as having a role in the regulation of iron absorption.[14–16] The release of iron from a ferri-transferrin-like complex would presumably involve a reductive mechanism[17] and it is conceivable that ascorbic acid (or other reducing agents) could influence absorption at this stage: studies in guinea-pigs would appear to indicate that it is the overall ascorbic acid status rather than the intraluminal concentration that modifies iron absorption.[18] Nevertheless, it is difficult to discount the evidence that iron enters the mucosal cell in the ferrous form[3,19] in which case a ferroxidase-like system must presumably be present in the mucosal cell before the iron can be trapped by a ferritin-like protein carrier molecule.[17] It is of course possible that ascorbic acid influences the absorption at two points and that the complete sequence is as follows: iron is transferred into the mucosal cell cytoplasm in the ferrous form and the intraluminal ascorbic acid will therefore enhance this movement. Within the mucosal cytoplasm a ferroxidase-like system converts the iron to the ferric form thereby enabling it to be bound to a carrier or acceptor molecule which effects its transfer to the serosal side where its liberation from the carrier molecule would require the presence of ascorbic acid.[17] A recent autoradiographic study has indicated that the intestinal absorption of iron is a two-phase process.[18]

Recent work on the relationship between age and absorption has indicated that the mechanism for iron absorption, when compared with that for lead and strontium, is not necessarily a specific one.[21] Is then the potentiating effect of ascorbic acid on iron absorption a specific example of a more general effect that operates for a number of other metals? Current evidence is insufficient to produce a satisfactory answer. Ascorbic acid in the intestinal lumen significantly depressed the absorption of ^{64}Cu in rats[22] and decreased the availability of copper to miniature swine.[23] On the other hand, *in vitro* studies have indicated that ascorbic acid potentiates the intestinal transport of manganese, cadmium and zinc[24] although the statement that 'The enhancing action of ascorbate on metal transport appears to be unrelated to its powerful reducing activity since dehydro-ascorbate has a similar effect'[24] must be challenged in the light of the finding that the gastro-intestinal mucosa is a potent source of a glutathione:dehydro-ascorbate oxidoreductase factor which rapidly converts dehydro-ascorbate back to ascorbate in the presence of glutathione.[25] High dietary intakes of ascorbic acid produced increased tissue concentrations of mercury in guinea-pigs given controlled dietary levels of mercuric mercury,[16] a finding which, if extrapolatable to the human situation, could be of considerable significance.

INTERACTIONS AT THE TISSUE LEVEL (a) IRON

There is considerable evidence relating iron mobilisation and storage to ascorbic acid status. It would appear that in both guinea-pigs and man the binding and transport of iron is influenced by ascorbic acid status. By converting iron into the ferrous state ascorbic acid increases its release from its bound and transport forms.[27–30] Wapnick *et al.* showed that seven days' therapy with ascorbic acid (500 mg/day) increased the desferrioxamine urinary iron excretion significantly in cases of Bantu siderosis.[31] Desferrioxamine chelates with iron and so increases its excretion in the urine; it chelates preferentially with the ferric form and Wapnick *et al.* suggest that reducing agents such as ascorbic acid may facilitate its action by releasing iron (Fe II) from storage forms to augment the pool of readily-available iron (Fe III). They point out that their results carry two implications (a) it would appear to be important to replete subjects with ascorbic acid before using desferrioxamine as a diagnostic aid for siderotic conditions and (b) alternative therapy with desferrioxamine (in place of repeated phlebotomies) may be more effective in siderotic subjects if adequate levels of ascorbic acid are maintained.[31]

A further finding is that the level of iron in the tissues influences the amount of ascorbic acid in the blood (and possibly in the tissues also); elevated tissue levels of iron are accompanied by low blood ascorbic acid concentrations and vice versa. Siderotic subjects have reduced leucocyte ascorbic acid levels[32] and the epidemiological pattern of scurvy in Johannesburg Bantu has been shown to be similar to that of severe iron overload.[33] In a separate study Jacobs *et al.* showed that patients with iron-deficiency anaemia had higher-than-average concentrations of leucocyte ascorbic acid whereas those with iron overload had reduced ascorbic acid levels.[34] Findings of this type are not entirely unexpected; any increase in the activity of the ferrous–ferric cycles associated with binding and release of iron would presumably result in an increased metabolic demand for ascorbic acid;[17] conversely a reduction in iron levels would have a 'sparing' effect on ascorbic acid. In addition to this there is some evidence that iron may exert a more-or-less direct influence on ascorbic acid metabolism by stimulating its oxidative breakdown to oxalic acid,[32,33] and it has been suggested that the low ascorbic acid levels found in siderotic subjects is a consequence of an iron-induced increase in its catabolism.[32,33] Experiments with guinea-pigs have given less conclusive results. Lipschitz *et al.* found that ascorbic acid deprivation altered the distribution pattern of storage iron and they suggested that ascorbic acid may be necessary, not only for the release of bound iron but also for the incorporation of transferrin-bound iron into hepatic ferritin.[30] Glover *et al.* using more realistic levels of ascorbic acid intake were unable to demonstrate any effect of iron overload or deficiency on vitamin C status in guinea-pigs.[18] This could well be a species difference and need not invalidate the main conclusions that in human subjects (a) dietary ascorbic acid status influences the uptake and subsequent metabolism of iron (b) changes in the iron status influence blood concentrations of ascorbic acid.

INTERACTIONS AT THE TISSUE LEVEL (b) OTHER METALS

It is of interest to consider to what extent similar relationships between ascorbic acid and other metals exist. It is possible that interactions of the type described are more or less specific in that they represent an evolutionary response to the peculiar status of iron as an essential nutrient; alternatively, they could be representative of a more general group of reactions common to a number of metals, and the behaviour of ascorbic acid *vis-à-vis* the uptake and metabolism of potentially toxic metals such as cadmium and mercury could in this respect be of considerable interest. In the case of mercury there has, for many years, been presumptive evidence of a relationship with vitamin C.

'Shun mercury as poison' was Kramer's advice to scorbutic patients according to George Budd, a London physician, in 1840. Budd himself claimed that in cases of scurvy '. . . mercury in every form should be religiously avoided (as) we have met with instances in which the scorbutic symptoms seemed to have been much aggravated by mercury, taken before the scurvy made its appearance'.[35]

Budd had a suitably cautious scientific approach and it is unlikely that he would have emphasised this apparent relationship between mercury and scurvy had he not a strong observational basis for doing so.[36] Currently, dietary intakes of ascorbic acid in Britain vary from the inadequate to the massive doses recommended by the 'mega-vitamin C therapists'.[37,38] Any relationship between the dietary intake of ascorbic acid and the metabolism of mercury could therefore be of some significance. Recent work with guinea-pigs showed that mercuric mercury significantly reduced the concentration of ascorbic acid in the brain, adrenals and spleen of animals given maintenance doses of the vitamin, a finding in agreement with Budd's observation that mercury exacerbated the scorbutic condition in man.

There is some evidence that ascorbic acid can protect an organism against the toxic action of some metals. An early report suggested that it would counteract the toxic action of arsenic compounds in man[39] and more recently Fox and Fry have shown that cadmium toxicity in the Japanese quail is decreased by ascorbic acid.[40,41] Large doses of ascorbic acid prevented the mercury-induced adrenal hypertrophy in guinea-pigs but did not prevent the changes induced by mercury in the weights of other organs.[26] Further studies in this field are indicated; it would be of value to measure the extent to which the general ascorbic acid status of an animal determines its susceptibility to mercury poisoning.

SULPHYDRYL INVOLVEMENTS

No discussion of the interactions between ascorbic acid and metals would be complete without a reference to sulphydryl (thiol) groups of proteins. The biological activity of many proteins is determined by their SH group content: alterations in the SH content may result in loss of enzyme activity, interference with membrane transport, etc.[43,44] Ascorbic acid, an easily diffusible biological reductant could, when present at the appropriate level, contribute to the maintenance of the integrity of SH groups. On the other hand, there is evidence that sulphydryl substances have a role in the preservation of tissue ascorbic acid by preventing its oxidation to dehydro-ascorbic acid[25,42] and any reduction in their biological activity could, indirectly, result in lowered tissue levels of ascorbic acid.

Sulphydryl groups are susceptible to attack by certain metals; in particular it is generally believed that many of the toxic effects of mercury result from a reaction of this type[45-48] although there are indications that it can also reduce enzyme activities by reacting with groups other than thiol ones; [47,49,50] Miller *et al.* were unable to find any change in liver and kidney SH content after treatment of chickens with methyl-mercuric chloride.[51] The use of different 'sulphydryl agents'[44] to study the distribution of thiol groups in guinea-pigs receiving different amounts of mercury and ascorbic acid would cast further light on this relationship.

GENERAL CONSIDERATIONS

The interactions between iron and ascorbic acid pose no nutritional problems provided that the intakes of both interactants approximate to normal levels. A high ratio of dietary iron: ascorbic acid, however, raises a special problem. In the case of interactions between ascorbic acid and non-nutrient metals, the nutritional implications are not as clearly definable. Lewin, reasoning essentially from theoretical considerations, has suggested that large dietary intakes of ascorbic acid could reduce the toxicity of certain metal ions by increasing their urinary excretion as ascorbic acid complexes.[13] This could well be so, but one should also bear in mind that certain tissues such as the brain have a high 'retention capacity' for ascorbic acid;[52] if this applies to ascorbic acid–metal complexes as well, then high ascorbic acid intakes would perhaps serve only to concentrate toxic metals in the brain and other tissues. Again, there is evidence that high intakes of ascorbic acid may actually increase the intestinal uptake of mercury and other metals[24,26]—possibly by maintaining the integrity of SH groups essential for their transport across the intestinal membrane.

On the other hand, there is evidence that at the tissue level itself, ascorbic acid concentration may be critical in influencing the degree of metal-induced toxicity (possibly by protecting SH groups essential for normal biological activity). Should one therefore advise a person in a 'high risk' mercury environment to increase or decrease his daily intake of ascorbic acid? Clearly there exist currently too many imponderables and too few clearly definable relationships to enable one to give a clear answer to this and other similar questions.

REFERENCES

1. Sasmal, N., Mukherjee, D., Kar, N. C. and Chatterjee, G. C. (1968). *Ind. J. Biochem.*, **5,** p. 123.
2. Sasmal, N., Kar, N. C., Mukherjee, D. and Chatterjee, G. C. (1968). *Biochem. J.*, **106,** p. 633.
3. Dowdle, E. B., Schachter, D. and Schenker, H. (1960). *Amer. J. Physiol.*, **198,** p. 609.
4. Chatterjee, G. C. (1967). In *The Vitamins*, Eds. W. H. Sebrell and R. S. Harris, 2nd edition, Vol. 1, Academic Press, p. 407.
5. Wiseman, G. (1964). *Absorption from the Intestine*, Academic Press, London and New York, p. 260.
6. Bothwell, T. H. (1968). *Br. J. Haematol.*, **14,** p. 453.
7. Bothwell, T. H. and Charlton, R. W. (1970). *Ann. Rev. Med.*, **21,** p. 145.
8. Forth, W. and Rummel, W. (1973). *Physiol. Review*, **53,** p. 724.
9. O'Brien, J. R. P. (1973). *Biochem. Soc. Trans.*, **1,** p. 70.
10. Pollack, S., Kaufman, R. M. and Crosby, W. H. (1964). *Blood*, **24,** p. 577.
11. Pollack, S., Kaufman, R., Crosby, W. H. and Butkiewicz, J. E. (1963). *Nature, Lond.*, **199,** p. 384.
12. Schade, S. G., Cohen, R. J. and Conrad, M. E. (1968). *New Engl. J. Med.*, **279,** p. 672.
13. Lewin, S. (1973). *Comp. Biochem. Physiol.*, **46B,** p. 427.
14. Levine, P. H., Levine, A. J. and Weintraub, L. R. (1972). *J. lab. clin. Med.*, **80,** p. 333.
15. Pollack, S., Campana, T. and Arcario, A. (1972). *J. lab. clin. Med.*, **80,** p. 322.
16. Sheehan, R. G. and Frenkel, E. P. (1972). *J. clin. Invest.*, **51,** p. 224.
17. Frieden, E. (1973). *Nutr. Rev.*, **31,** p. 41.
18. Glover, J. M., Jones, P. R., Greenman, D. A., Hughes, R. E. and Jacobs, A. (1972). *Br. J. exp. Path.*, **53,** p. 295.
19. Kuhn, I. N., Layrisse, M., Roche, M., Martinez, C. and Walker, R. B. (1968). *Amer. J. clin. Nutr.*, **21,** p. 1184.
20. Bedard, Y. C., Pinkerton, P. H. and Simon, G. T. (1971). *Blood*, **38,** p. 232.
21. Forbes, G. B. and Reina, J. C. (1972). *J. Nutr.*, **102,** p. 647.
22. Van Campen, D. and Gross, E. (1968). *J. Nutr.*, **95,** p. 617.
23. Voelkner, R. W., Jr. and Carlton, W. W. (1969). *Amer. J. vet. Res.*, **30,** p. 1825.
24. Sahagian, B. M. *et al.* (1967). *J. Nutr.*, **93,** p. 291.
25. Grimble, R. F. and Hughes, R. E. (1967). *Experientia*, **23,** p. 362.
26. Blackstone, S., Hurley, R. J. and Hughes, R. E. (1974). *Fd Cosmet. Toxicol.* (in press).
27. Mazur, A., Baez, S. and Shorr, E. (1955). *J. biol. Chem.*, **213,** p. 147.
28. Mazur, A., Green, S. and Carleton, A. (1960). *J. biol. Chem.*, **235,** p. 595.
29. Mazur, A. (1961). *Ann. N.Y. Acad. Sci.*, **92,** p. 223.

30. Lipschitz, D. A., Bothwell, T. H., Seftel, H. C., Wapnick, A. A. and Charlton, R. W. (1971). *Br. J. Haematol.*, **20,** p. 155.
31. Wapnick, A. A., Lynch, S. R., Charlton, R. W., Seftel, H. C. and Bothwell, T. H. (1969). *Br. J. Haemat.*, **17,** p. 563.
32. Wapnick, A. A., Lynch, S. R., Krawitz, P., Seftel, H. C., Charlton, R. W. and Bothwell, T. H. (1968). *Brit. med. J.*, **3,** p. 704.
33. Lynch, S. R., Seftel, H. C., Torrance, J. D., Charlton, R. W. and Bothwell, T. H. (1967). *Amer. J. clin. Nutr.*, **20,** p. 641.
34. Jacobs, A., Greenman, D., Owen, E. and Cavill, I. (1971). *J. clin. Path.*, **24,** p. 694.
35. Budd, G. (1840). In *The Library of Medicine,* Ed. Alexander Tweedie, Vol. V, London, p. 94.
36. Hughes, R. E. (1973). *George Budd in Med. Hist.*, **17,** p. 127.
37. Brook, M. (1972). *Nutritional Deficiencies in Modern Society,* Eds. A. N. Howard and I. M. Baird, London, p. 45.
38. Hughes, R. E. (1973). *Br. Nutr. Fdn Bull.*, **9,** p. 24.
39. Bundesen, H. N., Aron, H. C. S., Greenebaum, R. S., Farmer, C. J. and Abt, A. F. (1941). *J. Amer. med. Assn,* **117,** p. 1692.
40. Fox, M. R. S. and Fry, B. E. Jr. (1970). *Science,* **169,** p. 989.
41. Fox, M. R. S., Fry, B. E. Jr, Harland, B. F., Schertel, M. E. and Weeks, C. E. (1971). *J. Nutr.*, **101,** p. 1295.
42. Hughes, R. E. and Maton, S. C. (1968). *Br. J. Haemat.*, **14,** p. 247.
43. Erliz, D. and Leblanc, G. (1971). *J. Physiol.*, **214,** p. 327.
44. Motais, R. and Sola, F. (1973). *J. Physiol.*, **233,** p. 423.
45. Passow, H., Rothstein, A. and Clarkson, T. W. (1961). *Pharmac. Rev.*, **13,** p. 185.
46. Hughes, M. N. (1972). *The Inorganic Chemistry of Biological Processes,* Wiley, New York.
47. Webb, J. L. (1966). *Enzyme and Metabolic Inhibitors,* Vol. 2, Academic Press, New York and London.
48. Clarkson, T. W. (1972). *Biochem. J.*, **130,** p. 61.
49. Strominger, J. L. and Mapson, L. W. (1957). *Biochem. J.*, **66,** p. 567.
50. Dey, P. M. and Pridham, J. B. (1969). *Biochem. J.*, **115,** p. 47.
51. Miller, V. L., Bearse, G. E., Russell, T. S. and Csonka, E. (1969). *Poultry Sci.*, **48,** p. 1736.
52. Hughes, R. E. (1973). In *Molecular Structure and Function of Food Carbohydrate,* Eds. G. G. Birch and L. F. Green, Applied Science, London, p. 108.

DISCUSSION

Hornig: You stated that the brain releases ascorbic acid not at all easily. Does its release not follow, as in all other tissues, more or less first order kinetics? This would mean that less than about 5% of the original ascorbic acid would be present in the brain after 20 days on a vitamin-C deficient diet.

Hughes: The brain and the eye lens appear to differ from other organs in that the rate of release of ascorbic acid is much lower. After 14 days on a scorbutogenic diet the percentages of original ascorbic acid remaining in the adrenals, spleen, brain and eye lens are 4, 5, 24 and 27 respectively; after 20 days depletion the corresponding values are 1, 1, 15 and 18. (Hughes, R. E., Hurley, R. J. and Jones, P. R. (1971). *Br. J. Nutr.*, **26,** p. 433.)

Ford: In animals given a 1% solution of ascorbic acid instead of drinking water, was the effect on deposition of mercury perhaps non-specific—might the ingestion of some other acid have had a similar effect?

Hughes: We carried out a pilot test to determine the maximum level we could get without impairing the level to any great extent. We had hoped the vitamin C level in the other experiment would act as an index to toxicity, and we are still hoping we can use vitamin C in this.

Hornig: What is the method you have used to determine the ascorbic acid levels in the tissues of guinea-pigs?

Hughes: The 2,6-dichlorophenolindophenol photometric method with appropriate modifications of pH and timing to eliminate interfering substances. (Hughes, R. E. (1956). *Biochem. J.*, **64,** p. 203.)

Hornig: You stated that giving ascorbic acid as a 1% solution in the drinking water achieves tissue saturation. From our own studies we find that we must give at least two daily doses of 320 mg to achieve saturation.

Hughes: By giving the ascorbic acid in the drinking water we provide a daily intake of *ca.* 200–300 mg spread over 24 hr—thus approaching an infinite number of doses. This results in tissue saturation.

Lewin: Have you related the amount of vitamin C given, and the ascorbic acid in the blood, to the amount deposited in the tissues? This is a fundamental question. You can excrete increasingly large quantities. We should agree on some criteria here. Did the animals appear better or worse?

Hughes: We have not done anything on the distribution as between the tissues and the blood, because it was felt that the mercury level in the blood was critical.

Lewin: If precipitated, not saturated, it could do harm.

Hughes: I agree. We have not compared the different types of mercury in this context.

6

Vitamin C and Nitrosamine Formation

C. L. WALTERS

British Food Manufacturing Industries Research Association, Leatherhead, Surrey, England

ABSTRACT

Not only secondary but also tertiary amines, amine oxides and quaternary ammonium compounds can act as precursors to nitrosamines on reaction with a nitrite or another nitrosating agent. In most instances, the rate of nitrosamine production is dependent upon the nitrite concentration to an order of greater than one and thus the deprivation of nitrite through a competing pathway can bring about more than a concomitant reduction in the velocity of formation of nitrosamines.

The reaction at acid pH of ascorbic acid with nitrite, with the formation of nitric and nitrous oxides and nitrogen, is considered to cease as the pH approaches 6·0 *but nevertheless the production of nitric oxide has been observed even at this pH and the use of ascorbate in meat products is generally recognised as leading to lowered nitrite levels. However, the virtual elimination of nitrosamine production from nitrite and secondary amines which can occur in aqueous solution in the presence of ascorbic acid would not be anticipated solely on the basis of the reduced availability of nitrite. It has been suggested that nitrite forms a complex with ascorbic acid which renders the former unavailable for participation in the nitrosation of secondary amines but which continues to react as inorganic nitrite in analytical procedures.*

Not only can ascorbic acid prevent the formation of nitrosamines in vitro. *it also intervenes in the synthesis of N-nitroso compounds* in vivo *from ingested nitrite and secondary amines or amides and can prevent thereby their acute toxic, carcinogenic and teratogenic effects. A requirement of free enolic hydroxyl groups at carbon atoms* 2 *and* 3 *seems to have been established in that both dehydro-ascorbate and ascorbate-2-sulphate were inactive. The inclusion of ascorbate in cured meat products such as fried bacon and frankfurters is generally beneficial in reducing but not always eliminating the formation of volatile nitrosamines.*

Far less evidence is available for the action of ascorbic acid in vivo *on preformed nitrosamines than that for its effect on the synthesis of N-nitroso compounds from their component precursors. However, composite mixtures of ascorbate with ferrous salts and a chelating agent such as EDTA in the presence of oxygen can stimulate the action of liver microsomes in promoting the breakdown of nitrosamines, a process which is considered to commence with hydroxylation at a carbon atom and to proceed through the intermediate formation of the very reactive diazoalkanes.*

INTRODUCTION

The great majority of nitrosamines have been found to induce tumours in various sites when administered to experimental animals by a number of routes. One in particular, N-nitrosodiethylamine, has proved to be carcinogenic to no fewer than eleven different species,[1] ranging from fish to higher primates, so that man himself is unlikely to be exempt from its action.

Some nitrosamines are acutely toxic, particularly to individual species. It was, in fact, by this property that the formation of N-nitrosodimethylamine in fish meal heated with nitrite manifested itself in Norway, in that the concentration formed was sufficient to cause the deaths of sheep to which it was fed[2]. However, this represented an extreme case in that marine fish can contain large amounts of the precursor secondary amine involved and that massive levels of nitrite were employed. In addition to being carcinogenic, nitrosamines can also be mutagenic and teratogenic, inducing tumours and other malformations in the progeny of treated pregnant animals.

Not only is the formation of nitrosamines in foods and other environmental sources of potential concern, but the possibility remains of the synthesis of N-nitroso compounds *in vivo* in the gastro-intestinal tract from ingested nitrite or nitrate, with microbiological intervention in the latter case.

Vitamin C has already been widely used during curing, principally to promote a more effective utilisation of added nitrite in the formation of the cured meat pigment, nitrosylmyoglobin. In a scheme[3] postulated for the breakdown of nitrite with the formation of nitrosylmyoglobin in muscle tissue, ascorbic acid has been found to simulate the action of respiratory enzyme systems considered to be involved. Its effect in promoting the production of the cured meat haem pigment from oxymyoglobin and nitrite has also been demonstrated;[4] from kinetic studies, an intermediate complex between nitrite and ascorbic acid has been postulated.

The metabolism of dialkyl nitrosamines has been considered to proceed via hydroxylation at an α-carbon atom, followed by dehydration to form a diazoalkane or a carbonium ion which is responsible for alkylation of nucleic acids. Ascorbic acid systems containing a ferrous salt and EDTA have been found to act on preformed nitrosamines in the presence of oxygen in a similar manner,[5] leading thereby to their decomposition.

Formation of Nitrosamines and Nitrosamides

The synthesis of N-nitroso compounds is usually, but not always, accomplished through the interaction of nitrous acid with secondary amines or amides. The reaction involves undissociated nitrous acid and the free amino compound; maximal rates of nitrosation of a secondary amine occur therefore around pH 3, below which the reaction is restricted somewhat by an increasing tendency for salt formation on the part of the amine. Salt formation by secondary amides is far less marked and therefore nitrosation continues to increase with fall of pH below 3.

The rate of nitrosation of a secondary amine is generally considered to be dependent on the square of the nitrite concentration, although an exception was recorded in the cases of acetyltryptophan[6] and of sarcosine,[7] for which a linear dependence was observed, in like manner to the nitrosation of secondary amides. However, the extent of nitrosation of a secondary amine is markedly dependent on the basicity of the compound involved, there being a 185 000-fold difference between the rates of nitrosamine formation from piperazine, with a pK_a of 5·57, and piperidine (pK_a = 11·2) at the optimum pH and under equivalent conditions.[8] Obviously there is less tendency for salt formation on the part of the more weakly basic amine.

The catalysis of the nitrosation of secondary amines by a number of anions, and notably iodide and thiocyanate, has been reported.[9] The latter is known to occur physiologically and its concentration in the saliva is higher in heavy smokers. Simple aldehydes, and formaldehyde in particular, are capable of stimulating nitrosation markedly, particularly at pH values above neutrality[10] at which very little formation of nitrosamines would normally be anticipated.

Contrary to current popular belief, tertiary amines can react with nitrite to form nitrosamines. The nitrosation of tertiary amines is considered to proceed via the nitrosative cleavage of an alkyl group as an aldehyde, followed by the formation of the N-nitroso derivative of the resultant dialkylamine. The rate of formation of N-nitrosodimethylamine (DMN) from trimethylamine is reported to be[11,12] proportional to the cube of the nitrite concentration. This pathway is of particular potential importance in relation to the nitrosation in the stomach of complex tertiary amine drugs from which simple nitrosamines can be split off in the presence of nitrite; the maximum concentration of nitrite present in the stomachs of food consumers is, however, likely to be much less than those sometimes employed in model studies in this context. Amine oxides and quaternary ammonium compounds can also give rise to nitrosamines on reaction with nitrite.[11]

The Interactions of Ascorbate with Nitrite

The oxidation of nitrite by ascorbate has been studied extensively by Dahn *et al.*[13] who have found that the kinetics of the interaction are markedly dependent on the pH over the range 0–4. They postulated the formation of an intermediate complex between the two reactants which could be the alkyl nitrite of the C_3 hydroxyl group; this, they considered, would break down rapidly to nitric oxide and 'monodehydro-ascorbic acid'.

Using an ascorbate:nitrite molecular ratio of 100:1, the reaction observed over the pH range 3–5, with the predominant production of nitric oxide together with some nitrous oxide and nitrogen, ceased as the pH was raised further;[14] the oxygen uptake observed during the aerobic interaction was, however, far greater than that required for the oxidation of nitric oxide to nitrogen dioxide, and probably resulted from the catalytic oxidation of ascorbate. Nevertheless, the anaerobic incubation of 8·6 mM ascorbate with 62 mM nitrite at pH 6·0, a typical pH for a meat product, was found manometrically to produce an appreciable nitric oxide fraction.[15]

On the basis of kinetic studies of the role of ascorbate in promoting the formation of the cured meat pigment, nitrosylmyoglobin, from oxymyoglobin and nitrite, Fox and Thomson[4] have similarly come to the conclusion that an intermediate complex was formed between ascorbate and nitrite, which decomposes slowly to nitric oxide.

Ascorbate and Nitrosamine Formation *in vitro*

In simple aqueous buffer using an ascorbate: nitrite molar ratio of 2:1, the conversion of N-methylaniline into its N-nitroso derivative was found to be inhibited by 60% at pH 3 and by 45% at pH 4 (Table 1).[16] However, this amine is only weakly basic and it nitrosates so readily that it competes successfully with ascorbate for available nitrite. Both morpholine and piperazine are converted less readily into their N-nitroso derivatives than N-methylaniline and thus ascorbate was able to suppress completely their reactions with nitrite throughout the pH ranges of 1 or 2 to 4 at which they were tested. The blocking by ascorbate of the nitrosation of methylurea proved to be somewhat less effective at pH 1 and 2 than at pH 3, where it

TABLE 1

Effect of ascorbate on the nitrosation of amines and amides. (From Ref. 16, by permission.)

Base	Ratio ascorbate:nitrate	pH	Time (min)	Inhibition of nitrosation (%)
N-methylaniline	2:1	3	5	60
		4	10	45
Morpholine	2:1	2	20	98
		3	30	100
		4	30	100
Piperazine	2:1	1	60	98
		2	20	98
		3	10	99
		4	15	99
Methylurea	2:1	1	10	92
		2	10	87
		3	30	100
Dimethylamine	2:1	1	60	−80
		2	60	−7
		3	60	74
		4	60	97
	1·67:1	2	1 440	92
Oxytetracycline	1·5:1	2·8	1 320	100
		3·8	1 320	99·9

was complete. As a secondary amide, methylurea would be expected to react with nitrite less readily at pH 3 than at the lower pH values where it would be able to compete more effectively with ascorbate for nitrite.

The action of ascorbate on the formation of DMN from dimethylamine proved to be markedly dependent on pH. Using an ascorbate: nitrite ratio of 2:1 at pH 4, almost complete blocking occurred, which was partially maintained at pH 3. At pH 2, slight stimulation of nitrosation occurred and this was massively increased to 80% at pH 1. Throughout this series, however, the extent of reaction of the strongly basic dimethylamine was much smaller than that of the other less basic amines involved in the study. The blocking action of ascorbate on the reaction at pH 2 was greatly increased almost to completion on extending the period of incubation some 24-fold to 24 h, but no explanation has been advanced for this.

Oxytetracycline is a tertiary amine antibiotic which is subject to nitrosative cleavage to yield DMN. Over a long period, Mirvish *et al.*[16] reported virtually the complete prevention of the reaction at pH values of both 2·8 and 3·8. Since studies with other tertiary amines indicate that the primary reaction of nitrite with oxytetracycline would be to yield dimethylamine, it suggests that ascorbate reacts with nitrite sufficiently rapidly to prevent the formation of dimethylamine from oxytetracycline.

Ascorbate had no effect on the decomposition of the N-nitroso derivatives of methylurea and the amines listed in Table 1, 'indicating that its action in reducing or eliminating the formation of N-nitroso compounds is solely a consequence of its reaction with nitrite' according to Mirvish *et al.* The ascorbate ion is nitrosated 240 times more rapidly than ascorbic acid ($pK_a = 4{\cdot}29$) itself, due presumably to the greater nucleophilic activity of the anion. For this reason, ascorbate is more effective as a blocking agent at pH 3–4 than at pH 1–2, where the concentration of the ascorbate ion is much less.

These studies have been extended[17] to include the effect of ascorbate on the nitrosation of proline, hydroxyproline, sarcosine, histidine, citrulline and pipecolic acid, which are considered to be of biological origin, in a buffer of pH 3·0 (close to the optimum). Throughout, complete protection from nitrosation was achieved provided the ascorbate:nitrite ratio was unity or greater on a molecular basis. In order to block completely the nitrosation at pH 4 of morpholine, Fan and Tannenbaum[18] needed an ascorbate: nitrite ratio of at least 2:1; below this ratio, the formation of nitrosomorpholine was only partly inhibited. To explain their results, they have proposed that nitrite can react with ascorbic acid to become unavailable for nitrosation reactions. On the addition of the colour reagents for its determination, nitrite bound to ascorbate is released and becomes measurable.

However, Sen and Donaldson in Canada were able to claim[19] only partial blocking with ascorbic acid of the conversion of the anthelmintic amine piperazine at pH 2·0 in human gastric juice into mononitrosopiperazine (Fig. 1) using an ascorbate:nitrite ratio of approximately 16:1 on a molecular basis. Furthermore, ascorbic acid was found to stimulate the much smaller formation of dinitrosopiperazine, an observation which has not yet been explained.

Ascorbate and nitrosamine formation *in vivo*

N-nitrosoethylurea administered to pregnant animals has been found to be both teratogenic and carcinogenic to the offspring.[20,22] Similar results have been obtained when the precursors, ethylurea and nitrite, were given orally during the second half of pregnancy.[23–25]

After the administration to pregnant Wistar rats of ethylurea and nitrite, 60% of the progeny developed symptoms of hydrocephalus within 4–6 weeks, death occurring within 2–3 weeks further.[26] No closural defects were observed over one year, however, in the offspring of mothers treated concurrently with ascorbic acid along with the same amounts of ethylurea and nitrite.

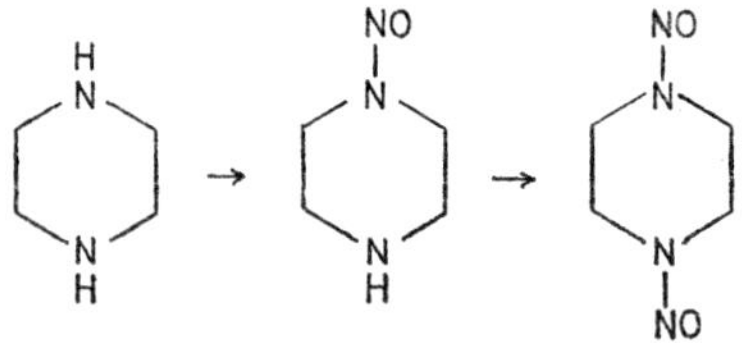

FIG. 1. Nitrosation of piperazine.

In a small scale experiment, complete protection had been provided at the time of reporting, by ascorbic acid administered to pregnant rats along with ethylurea and nitrite, from death due to induction of malignant tumours of neurogenic origin observed in a large proportion (72%) of the progeny in its absence. No protection was given, however, to animals treated with the intact nitrosamide; in fact, 20% more of the animals had died within 300 days after birth in the group treated with the nitrosamide and ascorbate, although the significance of this difference is doubtful.

The combined oral administration of nitrite with aminopyrine (Fig. 2) causes severe hepatic necrosis in the rat which is characterised by an elevation in serum glutamic pyruvate transaminase (SGPT). The hepatotoxicity presumably arose from the formation of N-nitroso-dimethylamine resulting from nitrosative cleavage of aminopyrine by nitrous acid in the rat stomach. It was found by Kamm *et al.*[27]

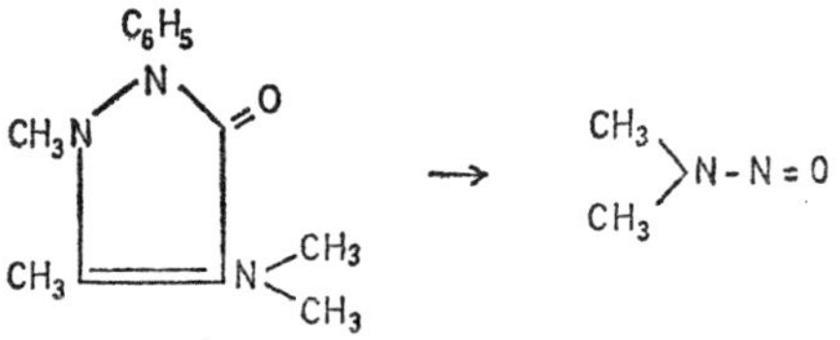

FIG. 2. Reaction of aminopyrine with nitrite.

that the concurrent administration of either sodium ascorbate or ascorbic acid completely blocked the rise in serum SGPT induced by the oral treatment of rats with sodium nitrite plus aminopyrine. Equimolar palmitoylascorbate proved to be somewhat less effective than ascorbate probably because of its limited solubility in aqueous media. Dehydro-ascorbate, in which the two enolic hydroxyl groups at C_2 and C_3 have been oxidised to carbonyl groups, and ascorbate-2-sulphate, afforded no protection. Protection was also given against the rise in SGPT resulting from the ingestion of dimethylamine and nitrite.

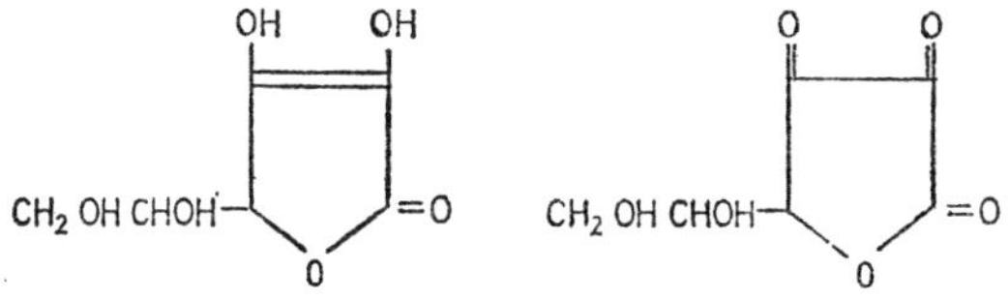

Ascorbic acid Dehydro-ascorbic acid

FIG. 3.

An oral dose of nitrite, followed by an intraperitoneal injection of aminopyrine did not lead to a rise in serum SGPT, as was obtained when both were administered orally. Furthermore intraperitoneal ascorbate did not protect animals from hepatotoxic effect resulting from the oral administration of the two precursors to DMN. These experiments demonstrate that the nitrosamine is formed from nitrite and aminopyrine in the rat's stomach and that ascorbate must be available in this organ to exert its protective action.

Following a single oral dose to rats of nitrite plus aminopyrine, the highest serum DMN level was found by Kamm *et al.*[27] about 30 min after the treatment. Sodium ascorbate administered concurrently with the nitrosamine precursors reduced the DMN concentration markedly and, at a level of 70 mg per kg body weight and above, the liberation of the nitrosamine into the serum was completely suppressed. For a 70 kg adult, the dosage required would approximate to 5 g on a body weight basis.

So far, these workers have not been able to decide unequivocally whether or not ascorbate has any effect on the actions of pre-formed nitrosamines administered to experimental animals. However, the concurrent administration of sodium ascorbate, together with oral N-nitrosodimethylamine, partially prevented the liver damage produced by the latter alone, suggesting a form of protection not involving competition for nitrous acid.[32]

Mirvish *et al.*[28] attempted to quantify the amount of ascorbate required to be administered in rat food together with 0·625% piperazine and 0·1% sodium nitrite in the drinking water to inhibit the production of lung adenomas. Tumour induction after 20 weeks of treatment and 10 weeks without treatment was suppressed to the extent of 91%, 69% and 37% through the inclusion in the food of 2·3%, 1·15% and 0·575% respectively of ascorbate, in comparison with the numbers of adenomas observed in the group fed piperazine plus nitrite without ascorbate. Furthermore, protection of a similar order was afforded when piperazine was replaced by morpholine.

Ascorbate and nitrosamine formation in foodstuffs

The formation of N-nitrosodimethylamine (DMN) in foodstuffs preserved with nitrite is sporadic and has not yet been associated with any technological or other parameter. Of a total of 197 such products, for instance, Panalaks *et al.*[29] detected DMN at the μg per kg level in 57.

Similarly, Fiddler *et al.*[30] at the Eastern Regional Research Center of the USDA detected DMN at levels of 11–84 parts in 10^9 in three out of 40 commercial frankfurters. Using the level of nitrite legally permitted in the USA in this type of product, namely 150 ppm, they were unable to prepare frankfurters experimentally containing measurable amounts of DMN. With approximately ten times the legal level of nitrite, however, they were able to produce frankfurters in their pilot plant which usually contained at least 10 μg DMN per kg. Table 2 records the results they obtained, using sodium

TABLE 2

Formation of DMN in frankfurters prepared with 1500 *ppm nitrite with and without sodium ascorbate and iso-ascorbate.* (*From Ref.* 30, *by permission.*)

Additive	Concentration	DMN (parts in 10^9) Processing time	
		2 hr	4 hr
None	—	11	22
Na ascorbate	(550 ppm)	0	7
	(5 500 ppm)	0	4
None	—	10	11
Na erythorbate	(550 ppm)	0	6
	(5 500 ppm)	0	0

ascorbate and sodium iso-ascorbate (erythorbate) added to frankfurters at the permitted level in the USA of 550 ppm and at ten times this amount.

Upon this basis, the use of ascorbate in cured meat products, albeit at relatively high levels, parallels its effects on nitrosation in simple aqueous solution. Iso-ascorbate also proved to be effective in this context.

The most critical situation in relation to the presence of nitrosamines in cured meat products is undoubtedly that of fried bacon, in which levels of N-nitrosopyrrolidine of 100 μg/kg can be produced after the severe cooking procedure typical of US practice. At a level of 330 mg/kg, the inclusion of ascorbic acid has not prevented the formation of this nitrosamine in fried bacon, though it probably reduces the amount produced. Experimental batches of bacon were also produced in the USA containing 1000 mg/kg ascorbic acid, above the limit of 550 mg/kg currently permitted. The higher concentrations appeared to lower further, but not to eliminate, the N-nitrosopyrrolidine level after frying. Unfortunately, however, it has proved impossible so far to obtain reproducible results between the laboratories involved due to variations, not the least of which is probably the frying procedure, which can be critical to the issue. The possibility exists that such high levels of ascorbate could interfere with the preservative and anti-botulinum effects of nitrite in bacon, though the issue is by no means clear as yet.

Effect of Ascorbate on Pre-formed Nitrosamines

The metabolism of simple dialkylnitrosamines has been considered to proceed via hydroxylation at an α-carbon atom, followed by dehydration leading to the formation of the very reactive corresponding diazoalkane or even a carbonium ion. In 1954 Udenfriend[31] reported a chemical system involving ascorbic acid which can simulate metabolic processes by hydroxylating aromatic compounds at the same electronegative sites where enzymic hydroxylation proceeds *in vivo*. The hydroxylation system includes ascorbic acid, the ferrous complex of EDTA and molecular oxygen. When applied to a series of 12 dialkylnitrosamines for a period of 6 hr, Preussmann[5] found that all were converted into compounds which could no longer be distilled in steam. The longer the carbon chain, the more rapidly the degradation proceeded, although straight chain compounds were decomposed more rapidly than branched. Only about 10% degradation persisted under nitrogen, probably due to the persistence of traces of oxygen. It was not possible to characterise the degradation products but simple hydrolysis to nitrite or nitrate was ruled out. This finding was extended to a number of heterocyclic nitrosamines with similar results.

As strong alkylating agents, diazoalkanes could themselves prove to be as carcinogenic as the nitrosamine of origin but it is unlikely that they would survive long in a biological matrix surrounded by carboxyl, amino, hydroxyl, sulphydryl and other reactive groups.

CONCLUSIONS

There is good evidence that ascorbic acid can inhibit or even block the nitrosation of secondary and tertiary amines and amides provided the ratio of the concentration of ascorbate and nitrite is sufficiently high. In general, its blocking action is least upon the formation of N-nitroso derivatives from weakly basic amines which are able to compete effectively with ascorbic acid for available nitrite. The action of ascorbate on the nitrosation of the strongly basic secondary amine dimethylamine is very markedly dependent upon pH, ranging from a stimulation at pH 1 to a strong inhibition at pH 4; no explanation has yet been advanced for this behaviour, although the kinetics of the reduction of nitrous acid by ascorbic acid has been found[13] to vary with pH over this range.

By virtue of its interference with the process of nitrosation, ascorbic acid has been effective in decreasing or eliminating the biological effects *in vivo* resulting from the concurrent administration of nitrite with amines and amides. It would appear that the enolic hydroxyl groups at carbon atoms 2 and 3 are necessary for its protective role. Little evidence has been obtained of protection against the actions of pre-formed nitrosamines.

Ascorbic acid is also active in reducing but not necessarily eliminating nitrosamine formation in cured meat products either produced with nitrite levels greatly in excess of the legally permitted content or which are subjected to strong heat processing.

Whilst ascorbic acid itself is not considered to have an action on pre-formed nitrosamines, systems including also ferrous ions and EDTA can bring about their decomposition to non-steam volatile products in the presence of oxygen.

REFERENCES

1. Schmähl, D. and Osswald, H. (1967). *Experientia,* **23,** p. 497.
2. Sakshaug, J., Sognen, E., Hansen, M. A. and Koppang, N. (1965). *Nature, Lond.,* **206,** p. 1261.
3. Walters, C. L., Casselden, R. J. and Taylor, A. McM. (1967). *Biochim. Biophys. Acta,* **143,** p. 310.
4. Fox, J. B. and Thomson, L. S. (1963). *Biochemistry,* **2,** p. 465.
5. Preussmann, R. (1964). *Arzneimittel Forsch.,* **14,** p. 769.
6. Kurosky, A. and Hofmann, T. (1972). *Canad. J. Biochem.,* **50,** p. 1282.
7. Friedman, M. A. (1972). *Bull. Environmental Contam. & Toxicol.,* **8,** p. 375.
8. Mirvish, S. S. (1972). *N-Nitroso Compounds—Analysis and Formation,* published by IARC, Lyon, France, p. 104.
9. Boyland, E., Nice, E. and Williams, K. (1971). *Food Cosmet. Toxicol.,* **9,** p. 1.
10. Keefer, L. K. and Roller, P. P. (1973). *Science,* **181,** p. 1245.
11. Fiddler, W., Pensabene, J. W., Doerr, R. C. and Wasserman, A. E. (1972). *Nature, Lond.,* **236,** p. 307.
12. Schweinsberg, F. and Sander, J. (1972). *Hoppe-Seyler's Z. physiol. Chem.,* **353,** p. 1671.
13. Dahn, H., Loewe, L. and Bunton, C. A. (1960). *Helv. Chim. Acta, XLIII,* p. 320.
14. Evans, H. J. and McAuliffe. (1956). *A Symposium on Inorganic Nitrogen Metabolism,* Ed. W. D. McElroy and B. Glass, The Johns Hopkins Press, Baltimore, p. 189.
15. Walters, C. L. and Taylor, A. McM. (1964). *Biochim. Biophys Acta.,* **86,** p. 448.
16. Mirvish, S. S., Wallcave, L., Eagen, M. and Shubik, P. (1972). *Science,* **177,** p. 65.
17. Walters, C. L. and Cave, J. M. (1972). Unpublished.
18. Fan, T. Y. and Tannenbaum, S. R. (1973). *J. Food Sci.,* **38,** p. 1067.
19. Sen, N. P. and Donaldson, B. (1973). *Third Meeting on the Analysis and Formation of N-Nitroso Compounds,* IARC, Lyon, France, to be published by WHO.
20. Ivankovic, S. and Druckrey, H. (1968). *Z. Krebsforsch.,* **71,** p. 320.
21. Druckrey, H., Ivankovic, S. and Preussmann, R. (1966). *Nature, Lond.,* **210,** p. 1378.
22. Druckrey, H., Preussmann, R., Ivankovic, S. and Schmähl, D. (1967). *Z. Krebsforsch.,* **69,** p. 103.
23. Ivankovic, S. and Preussmann, R. (1970). *Naturwiss.,* **57,** p. 460.
24. Osske, G., Warzok, R. and Schneider, J. (1972). *Arch. Geschwulst-Forsch.,* **40,** p. 244.
25. Alexandrov, V. A. and Jänisch, W. (1971). *Experientia,* **27,** p. 538.
26. Preussmann, R., Ivankovic, S., Schmähl, D. and Zeller, J. (1973). *Third Meeting on the Analysis and Formation of N-Nitroso Compounds,* IARC, Lyon, France, to be published by WHO.

27. Kamm, J. J., Dashman, T., Conney, A. H. and Burns, J. J. (1973). *Proc. Nat. Acad. Sci. USA,* **70,** p. 747.
28. Mirvish, S. S., Cardesa, A., Wallcave, L. and Shubik, P. (1973). *Proc. Amer. Assoc. Cancer Res.,* **14,** p. 102.
29. Panalaks, T., Iyengar, J. R. and Sen, N. P. (1972). *JAOAC,* **56,** p. 621.
30. Fiddler, W., Pensabene, J. W., Pistrowski, E. G., Doerr, R. C. and Wasserman, A. E. (1973). *J. Food Sci.,* **38,** p. 1084.
31. Udenfriend, S., Clark, C. T., Axelrod, J. and Brodie, B. B. (1954). *J. Biol. Chem.,* **208,** p. 731.
32. Kamm, J. J., Dashman, T., Conney, A. H. and Burns, J. J. (1973). *Proc. Nat. Acad. Sci., USA,* **70,** p. 747.

DISCUSSION

Wilson: Is there any evidence that nitrosamines can produce tumour growths in human beings?
Walters: No, I think not. I have heard of no epidemiological studies that have given rise to any concern here. There were a very small number of people who were subjected to relatively high concentrations of nitrosamines before their hazards were recognised, but I gather that it was very difficult to follow them up—the numbers involved were too small to give you an authoritative statement.

On the other hand, the metabolism of nitrosamines in human liver is virtually the same as that in affected species, so if that is anything to go by then human beings would be affected. But certainly the dosage involved is just not recognised.
Spencer: Is there any indication as to whether the use of ascorbic acid in cured meats affects in any way the function of nitrites in these products?
Walters: Overall, it has been considered to be a beneficial effect, though with some reservations as a result of recent work that has been done. So far as the antibacterial action of nitrite is concerned, the general consensus of opinion is that it neither detracts from nor enhances that property of nitrite.

7

Recent Advances in Vitamin C Metabolism

D. HORNIG

Department of Vitamin and Nutritional Research,
F. Hoffmann-La Roche & Co. Ltd, Basle, Switzerland

ABSTRACT

The recent isolation of ascorbic acid 2-sulphate from the urine of animals and humans as well as its presence in substantial amounts in tissues of rats, fish and brine shrimp cysts evoked the question whether so far unidentified metabolites of ascorbic acid may contribute to the elucidation of the mode of action of vitamin C.

The role of the biologically inactive ascorbic acid derivative, ascorbic acid 2-sulphate, as an in vivo *sulphating agent has been investigated in female rats. The faecal excretion of cholesterol and cholesterol sulphate after an intravenous application of a* ^{14}C*-labelled dose of cholesterol was unaffected by ascorbic acid 2-sulphate treatment. Also the accumulation of* ^{14}C*-radioactivity by various tissues occurred to a comparable extent.*

The metabolic fate of ascorbic acid 2-(^{35}S*) sulphate was followed in guinea-pigs and rats. After oral application to guinea-pigs, about 10–12% of the dose appeared in the urine during the first 24 hr. Thereof, 30–40% was identified as unchanged ascorbic acid 2-sulphate and about 60% as an even more polar metabolite. Rats excreted a lower percentage of the orally administered ascorbic acid 2-sulphate and the polar metabolite. The guinea-pig faeces contained about 5% of the given* ^{35}S*-labelled material (24 hr), mainly as ascorbic acid 2-sulphate (90%) and cholesterol sulphate, whereas in the rat faecal excretion was a major pathway for the elimination of label after oral dosage (30%, thereof 30% as the polar metabolite; 24 hr). These observations suggest the existence of a species-specific difference in the metabolism of ascorbic acid 2-sulphate. Results obtained with ascorbic acid 2-sulphate-6(*$^{3}H_2$*) given orally to rats indicate the unknown polar metabolite to be an oxidised compound, possibly a 6-carboxyl-derivative of ascorbic acid 2-sulphate.*

Studies on the tissue distribution of ascorbic acid 2-(^{35}S*) sulphate in guinea-pigs and rats by means of dissection and whole-body autoradiography indicated a poor accumulation by the various tissues. The pituitary and adrenal glands contained only minute (*^{35}S*) radioactivity. Possible implications of ascorbic acid 2-sulphate with regard to its distribution in the tissues will be discussed.*

INTRODUCTION

The metabolism of vitamin C (L-ascorbic acid) has been extensively investigated in humans and especially in animals using the radioactivity labelled compound. However, the knowledge of the mode of action of ascorbic acid (AA) is still far from complete. The recent isolation of ascorbic acid 2-sulphate (AAS) from the urine of animals as well as of humans[1] might give reason to believe that the knowledge of the metabolic function of AAS and other so far unidentified metabolites of AA could be pertinent to the elucidation of the physiological role of AA in humans as well as in animals. AAS was also isolated from brine shrimp cysts,[2,3] was demonstrated in rat organs,[4] and was suggested as a metabolite of ascorbic acid occurring in the bile[5] and lymph[6] of rats. The sulphated form of AA is suggested to represent an additional pool as a possible storage form for AA.[7]

Metabolic Role of Ascorbic Acid 2-Sulphate

Verlangieri and Mumma[8] reported an approximately 50-fold elevation of the (^{35}S)labelled cholesterol sulphate excretion in the faeces of rats after intracardial injection of (^{35}S)AAS in comparison to injection of $^{35}SO_4^{2-}$. Administration of AA together with $^{35}SO_4^{2-}$ caused a two-fold excretion of cholesterol (^{35}S)sulphate. These data were thought to be evidence for the existence of a sulphate transfer from AAS onto cholesterol under formation of cholesterol sulphate. These authors, therefore, postulated that the hypocholesterolemic effect of AA may be due to the intermediate synthesis of AAS.[8]

We have attempted to reproduce these findings but were not able to confirm the significant enhancement of the cholesterol sulphate excretion using the intravenous route for administration of the (^{35}S)AAS. We could, however, demonstrate that a limited part of the sulphate moiety of AAS is transferred onto cholesterol. Table 1 gives the figures for the total urinary and faecal excretion of (^{35}S)radioactivity during the first 72 hr following injection. Approximately 90% of the administered inorganic $^{35}SO_4^{2-}$ was found to be excreted in the urine, mainly in unaltered form (80%) and as cholesterol sulphate (15%) as judged from TL chromatography and electrophoresis. Faecal excretion amounted to approximately 0·6%. Co-chromatography of the chloroform–methanol extract

TABLE 1

Total urinary and faecal excretion of (^{35}S)radioactivity during the first 72 hr after injection of either 18·7 μC SO_4^{2-}, or 16·8 μC SO_4^{2-} and 5 mg ascorbic acid, or 20·59 μC ascorbic acid 2-(^{35}S) sulphate dipotassium salt. The figures are expressed as per cent of the administered labelled dose

Administered compound	Per cent of administered radioactivity Urine			Faeces		
$^{35}SO_4^{2-}$	91·7	89·2		0·61	0·64	
$^{35}SO_4^{2-}$ and ascorbic acid	38·3	32·8	37·3	0·21	0·18	0·21
Ascorbic acid 2-(^{35}S) sulphate	63·8	55·4	65·4		0·30[a]	

[a] Mean of 3 rats.

together with cholesterol sulphate indicated that this compound accounted almost exclusively for the faecal (^{35}S)radioactivity.

Only an average of 36·1% was excreted with the urine when $^{35}SO_4^{2-}$ was injected together with AA, and 0·2% appeared in the faeces. Approximately 90% of the excreted material in the urine was unaltered $^{35}SO_4^{2-}$. 80–90% of the (^{35}S)labelled material in the faeces was identified by chromatography as (^{35}S)cholesterol sulphate. The other part was even more unpolar than cholesterol sulphate and was not further identified.

After injection of (^{35}S)AAS approximately 40% remained in the animal body at 72 hr. The (^{35}S)labelled material excreted in the urine (61·5%) was identified as unaltered (^{35}S)AAS (90%), only traces of labelled cholesterol sulphate (less than 2%) were observed. An average of 0·3% was excreted during the first 72 hr after administration with the faeces. Thereof, 94% was identified as (^{35}S)labelled cholesterol sulphate, about 6% was in a more unpolar fraction.

On the other hand, our results are not in accordance with the finding by Verlangieri and Mumma, since these authors have reported an almost fifty-fold enhancement in the formation of labelled cholesterol sulphate (Table 2). It seems uncertain that there exists a difference in the enzyme system[9] necessary for splitting off and transferring the sulphate moiety in the different rat strains used.

Similar results were obtained in a study on the effect of AAS on the faecal excretion of exogenous cholesterol in rats.[10] The animals were administered intravenously AAS (0, 10 or 20 mg/day) or sodium ascorbate (33 mg/day) over 5 days and at the last day a single dose of (4-^{14}C)cholesterol. No stimulatory effect of AAS or AA was

TABLE 2

Comparison of (^{35}S)cholesterol sulphate excreted in the faeces after intravenous injection related to administered (^{35}S)dose. Figures taken from Verlangieri and Mumma[8]

Administered compound	Per cent	
	Own results[a]	Results from (8)[b]
$^{35}SO_4^{2-}$	0·62	0·16
$^{35}SO_4^{2-}$ and ascorbic acid	0·16–0·18	0·25
Ascorbic acid 2-(^{35}S)sulphate	0·28	7·1

[a] Excretion during first 72 hr.
[b] Excretion during first 45 hr.

observed during the first 72 hr after the last injection either on the faecal excretion of (^{14}C)cholesterol or on that of (^{14}C)cholesterol sulphate. The composition of the radioactive material in the faeces was unaffected (cholesterol 90%; cholesterol sulphate 6%) and no enhancement of the cholesterol sulphate excretion could be observed (Table 3).

Our results do not favour the hypothesis that a hypocholesterolemic effect of AA may be due to the intermediate formation of AAS.

TABLE 3

Percentage composition of the radioactive material excreted in the faeces during the first 72 hr after intravenous injection of 22·8 μC (4-^{14}C)cholesterol (159 μg) to female rats which were treated for 5 days with a daily injection of ascorbic acid 2-sulphate dipotassium salt (10 or 20 mg, $n = 7$) in 0·15M Na^+-phosphate buffer (pH 7·4) or with buffer. For controls 2 animals were injected either with 33 mg Na^+-ascorbate in buffer or with 10 mg K_2SO_4 in buffer. The radioactive material was extracted from the faeces with chloroform/methanol and the extracts were chromatographed on silica gel-coated plates

Administered compound (mg/day)		Per cent of excreted radioactivity		
		Cholesterol	Cholesterol sulphate	Polar fraction
Na^+-ascorbate	33	90·0	6·1	3·9
K_2SO_4	10	93·8	3·9	2·3
Buffer		89·9	6·2	3·9
Ascorbic acid 2-sulphate	10	89·7	6·3	4·0
	20	90·1	6·4	3·5

Furthermore, AAS cannot be considered as a hypocholesterolemic agent *per se*.

It has been postulated by Ginter[11] that AA controls the transformation of cholesterol to bile acids in guinea-pigs. We have, therefore, studied a possible effect of AAS on the biliary appearance of (^{14}C)radioactivity, derived from administration of (4-^{14}C)-cholesterol, in bile-duct cannulated rats during the first 72 hr after dosage of the label. The animals received daily over 5 days either 0, 50 or 200 mg of AAS in saline by gastric intubation. On day 6 a single dose of 30 μC (4-^{14}C)cholesterol was given by the same route after bile-duct cannulation. The total excretion of (^{14}C)radioactivity during the first 72 hr amounted to 33–35% of the administered dose. From Table 4 it is seen that the excretion of the unpolar

TABLE 4

Biliary excretion of radioactive material during the first 72 *hr after gastric intubation of* (4-^{14}C)*cholesterol* (30 μC; 210 μg) *following bile-duct cannulation in female rats. The animals had received prior to cannulation and administration of label over a period of* 5 *days* 0, 50 *or* 200 *mg AAS per day in saline. The amount of the unpolar and polar fractions excreted with the bile were determined by thin-layer chromatography on silica-gel-coated plates* (*benzene–methylethylketone–ethanol–water*, 30:30:30:10; *v*/*v*)

	(^{14}C)Radioactivity (dpm × 10^{-6})	
Ascorbic acid 2-sulphate (mg/day)	Unpolar fraction	Polar fraction
0	2·27	18·92
50	3·00	18·17
200	3·69	18·45

fraction (mainly cholesterol) in the bile was only slightly enhanced by increasing intakes of AAS, but the polar fraction (bile acids) remained unaltered. These findings, however, are evidence against the participation with regard to a regulatory role of AAS in the catabolism of cholesterol to bile acids.

Metabolism of Ascorbic Acid 2-Sulphate

The urinary as well as the faecal excretion of (^{35}S)labelled AAS has been studied in guinea-pigs and rats after oral and intravenous administration. The results are summarised in Table 5. In the guinea-pig about 10% of an oral dose was excreted in the urine during the first 8 hr and less than 5% appeared in the faeces. The observation that the rat excreted approximately 30–40% of the orally administered dose in the faeces, but comparable quantities with the urine indicates

TABLE 5

Percentage of urinary and faecal excretion of radioactive material following oral or intravenous (intracardial) administration of ascorbic acid 2-(^{35}S)-sulphate to guinea-pigs and rats. Urine was counted directly for radioactivity. The radioactive material in the faeces was counted after extraction with water and chloroform/methanol. The (^{35}S)radioactivity has been corrected for decay

Animal	Route	Ascorbic acid 2-(^{35}S)sulphate		Per cent of administered dose		Time after dosage (hr)
		μC	μg	Urine	Faeces	
Guinea-pig	oral	14·49	630	1·5	—	2
	oral	14·64	637	11·4	1·6	8
	oral	167·60	7 290	—	6·6	8
Rat	oral	23·00	1 000	5·8	25·5	8
	oral	167·60	7 290	8·1	46·6	8
	oral	11·20	487	18·3	33·1	24
	oral	11·20	487	11·8	27·2	24
Guinea-pig	i.c.	167·60	7 290	63·7	—	0·67
	i.c.	167·60	7 290	58·9	—	0·67
	i.c.	14·49	630	53·5	—	2
Rat	i.v.	14·63	636	60·1	—	1
	i.v.	13·30	578	37·6	—	8
	i.v.	13·30	578	37·2	—	8
	i.v.	13·30	578	19·1	0·15	24
	i.v.	13·30	578	27·4	0·01	24
	i.v.	25·60	1 113	66·2	0·3	72
	i.v.	25·60	1 113	57·4	0·3	72
	i.v.	25·60	1 113	67·8	0·3	72

the existence of a species-specific difference in the metabolism of AAS.

After intravenous injection of (^{35}S)AAS to the rat about 30–40% was excreted with the urine whereas less than 1% could be found in the faeces. More than 60% was eliminated by the urinary pathway after 72 hr. A rapid elimination of (^{35}S)radioactivity was also observed following intracardial injection of (^{35}S)AAS to guinea-pigs. Already 2 hr after injection 50–60% of the administered dose was recovered from the urine.

Studies have also been undertaken with (6-$^{3}H_2$)AAS in rats. After oral or intravenous administration urinary excretion of (^{3}H)radioactivity amounted to 4·2% or 63·0% (48 hr), faecal excretion was 19·7% or 36·7% of the dose, respectively.

In general, the results indicate a poor uptake of AAS by the tissues and a low capacity for re-absorption of AAS by the renal tubuli.

TABLE 6

Electrophoretic and thin-layer chromatographic data of the labelled material excreted in the urine of rats and guinea-pigs after oral or intravenous administration of ascorbic acid 2-(^{35}*S*)*sulphate. Solvents were as follows: paper electrophoresis:* 0·15*M ammonium acetate buffer, pH* 5·8*; Whatman* 3 *MM or Schleicher and Schull* 2043*a. Thin-layer chromatography (silica gel* F_{254}*): solvent I acetone–methanol–benzene–glacial acetic acid,* 5:20:20:5, *v/v; solvent II chloroform–methanol–water–glacial acetic acid,* 65:50:15:2, *v/v; (cellulose F) solvent III* 0·04*M ammonium formate, pH* 4·3-2-*propanol,* 50:40, *v/v*

	Electrophoresis $R_{picrate}$	Chromatography R_f-values I	II	III
Guinea-pig, *i.c.*				
(^{35}S)activity	2·03	0·53	0·16	0·59
Ascorbic acid 2-sulphate	2·03	0·53	0·16	0·59
Rat, *i.v.*				
(^{35}S)activity	2·45	0·60	0·31	0·64
Ascorbic acid 2-sulphate	2·45	0·60	0·30	0·65
Guinea-pig, *oral*				
(^{35}S)activity	2·43		0·18	0·59
	3·60		0·18	
Ascorbic acid 2-sulphate	2·43		0·17	0·59
Rat, *oral*				
(^{35}S)activity	2·24	0·56		
	3·28	0·08		
Ascorbic acid 2-sulphate	2·21	0·58		

Identification of Urinary and Faecal Metabolic Products

Attempts have been made to characterise the (^{35}S)labelled compound(s) excreted by these pathways, by means of high-voltage paper-electrophoresis and thin-layer chromatographic procedure. After intravenous administration the urinary material was almost exclusively identified as unchanged (^{35}S)AAS, whereas cholesterol sulphate accounted for the radioactivity excreted in the faeces. An even more polar compound than AAS amounting to about 60%, and unchanged AAS (30–40%), and an unpolar labelled substance (5%) were separated from the urine of guinea-pigs and rats. The polar metabolite of AAS was also observed in the faeces to about 15–20% of the excreted radioactive material. The R_f values for the labelled compounds excreted in the urine are given in Table 6.

The possibility that this polar compound is produced by the intestinal micro-organisms was ruled out by treating rats with aureomycin before and during the experiment. However, no changes in the excretion pattern occurred.

The more polar compound was thought to be an oxidised product of AAS, possibly in the 6-position. This was investigated after oral administration of (6-$^3H_2$)ascorbic acid 2-sulphate. In the rat less than 1% appeared in the urine (80% thereof as unchanged AAS; 20% as water). The polar metabolite was not observed by electrophoresis and thin-layer criteria. Also in the faeces (approximately

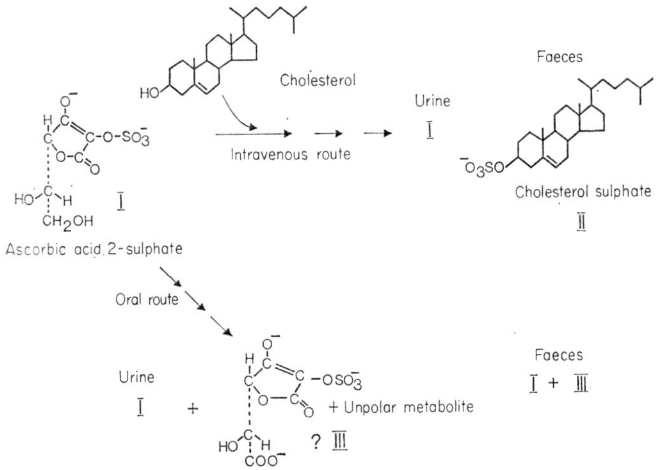

FIG. 1. 6-carboxyl-6-desoxy-ascorbic acid 2-sulphate.

20% of the dose excreted during the first 48 hr), the polar metabolite was not excreted (80–90% as unchanged AAS, about 10% as a relatively unpolar compound). The findings with the ^{3}H-labelled AAS would fit the hypothesis that the more polar metabolite might be a 6-carboxyl derivative of AAS.

The overall metabolism of AAS is given in scheme in Fig. 1.

Tissue Distribution after Administration of Ascorbic Acid 2-Sulphate

As indicated by the rapid elimination rates via urinary or faecal pathways (*see* above) the uptake of (^{35}S)radioactivity derived from (^{35}S)AAS by the tissues was found to be only minute. It seems to be noteworthy that almost no uptake of (^{35}S)labelled material by the adrenal and pituitary glands was observed. Most of the label was taken up by the kidneys and the liver. A difference between the rat and the guinea-pig is indicated in the uptake by the spleen and the testes, suggesting the rat to have a higher capacity to accumulate radioactive material. A summary of the obtained figures is presented in Table 7.

Results from whole-body autoradiographic studies have further shown an accumulation of the (^{35}S)radioactivity in the skin after either route of administration in both animal species. Also the presence of minute concentrations of labelled material was seen in the intervertebral discs, in the bone marrow and periosteum, and on the surface of the tongue. This would indicate that AAS *per se* or the SO_4^{2-}-moiety only is incorporated into polysaccharides. A comparable distribution of radioactivity has been observed after administration of $^{35}SO_4^{2-}$ to rats (Rietz, personal communication). The assumption of an AAS-bodypool as a storage form of AA[7] would imply that only a limited part of the AAS released from the bodypool can be converted, possibly under regulation of ascorbate sulphohydrolase,[9] to inorganic sulphate which then is utilised in the formation of the sulphated polysaccharides. On the other hand, the incubation of cultured human fibroblasts with (^{35}S)AAS resulted in insignificant incorporation of (^{35}S)radioactivity in acid muco-polysaccharides but in a compound more highly labelled than (^{35}S)AAS.[12]

The limited uptake of AAS by the tissues after either oral or intravenous administration, the rapid elimination of mostly unaltered AAS with urine and faeces and the indicated minute conversion of AAS to inorganic sulphate indicate only a low potency of AAS to

TABLE 7

Distribution of (^{35}S)radioactivity (dpm/total organ; rounded figures) after either oral or parenteral administration of ascorbic acid 2-(^{35}S)sulphate to guinea-pigs and rats. After anaesthesia with Penthrane[R] the abdomen of the animal was opened and it was sacrificed by bleeding caused by a cut of the portal vein. The organs were rapidly dissected and processed for counting of radioactivity. Solubilisation of the tissues was achieved by 1N NaOH. Insta-Gel[R] was used as counting solution

	Total radioactivity (dpm × 10^{-2})										
	Guinea-pig[a]		Guinea-pig[c]			Rat[a]			Rat[c]		
Administered dose (μC)	14·49		15·89	14·16	14·49	11·21			13·3		
Time after dosage (hr)	2	8	1	2	8	2	8	24	2	8	24
Lungs	113	410	1 146	460	63	24	546	85	96	28	17
Liver	458	2 540	2 897	2 630	952	289	5 643	765	2 550	355	125
Kidneys	2 069	1 040	510	3 910	120	150	1 684	301	500	59	40
Adrenal glands	6	20	52	13	2	1	20	3	2	1	0
Spleen	16	49	108	31	10	15	251	51	32	17	10
Testes	13	39	268	49	5	21	786	115	108	45	20
Brain	28	96	164	95	78	13	131	38	28	13	13

[a] Oral; [b] intracardial; [c] intravenous.

relieve scurvy. This view has been confirmed in guinea-pigs using oral as well as parenteral administration of AAS[13] (Weiser, unpublished results).

REFERENCES

1. Baker, E. M., Hammer, D. C., March, S. C., Tolbert, B. M. and Canham, J. E. (1971). *Science,* **173,** p. 826.
2. Mead, C. G. and Finamore, F. J. (1969). *Biochemistry,* **8,** p. 2652.
3. Bond, A. D., McClelland, B. W., Einstein, J. R. and Finamore, F. J. (1972). *Arch. Biochem. Biophys.,* **153,** p. 207.
4. Mumma, R. O. and Verlangieri, A. J. (1972). *Biochim. Biophys. Acta,* **273,** p. 249.
5. Hornig, D., Gallo-Torres, H. E. and Weiser, H. (1973). *Biochim. Biophys. Acta,* **320,** p. 549.
6. Hornig, D. and Gallo-Torres, H. E. (1972). *Scand. J. clin. Lab. Invest.,* **29,** Suppl. 126, abstr. 31–4.
7. Baker, E. M., Kennedy, J. E., Tolbert, B. M. and Canham, J. E. (1972). *Fed. Proc.,* **31,** p. 705, abstr. 2760.
8. Verlangieri, A. J. and Mumma, R. O. (1973). *Atherosclerosis,* **17,** p. 37.
9. Tolbert, B. M., Bullen, W. W., Downing, M. and Baker, E. M. (1973). *Fed. Proc.,* **32,** p. 931, abstr. 4007.
10. Hornig, D., Weber, F. and Wiss, O. (1974). *Z. klin. Chem. Biochem.,* in press.
11. Ginter, E. (1973). *Science,* **179,** p. 703.
12. Bond, A. D. (1973). *Fed. Proc.,* **32,** p. 932, abstr. 4011.
13. Campeau, J. D., March, S. C. and Tolbert, B. M. (1973). *Fed. Proc.,* **32,** p. 931, abstr. 4008.

DISCUSSION

Chairman: I should be interested to know, Dr Hornig, if ascorbic acid sulphate reacts in the normal assay procedure as ascorbic acid?

Hornig: By the normal methods you measure ascorbic acid sulphate together with ascorbic acid. I cannot recall the oxidant—do you know, Myron, is it iodine or is it charcoal?

Brin: If you use bromine to convert ascorbic acid to dehydro-ascorbic acid, then you determine both simultaneously by the sum, but if you use charcoal, only ascorbic is converted to dehydro, and the sulphate is untouched.

Chairman: My point was that one could miss the presence of ascorbic acid sulphate in a normal assay.

Hornig: This probably applies mostly to meat products.
Degkwitz: The redox potential for semidehydro-ascorbic acid to ascorbic acid is given as +0·34 V, which is rather high, and this is why I doubt that this vitamin reacts mainly as an electron donor. Do you know if the corresponding redox potential of the AA sulphate is different?
Hornig: No.
Fairhurst: One of the symptoms shown by the scorbutic animal is a depressed sulphation of chondroitin. Dr Hornig has suggested that ascorbic acid sulphate may be involved in the sulphation of polysaccharides. Has he made any specific studies on the sulphation of mucopolysaccharides, such as chondroitin sulphate, after administration of ascorbic acid sulphate?
Hornig: No, not yet.
Brin: Dr Shapiro in our group has undertaken to study ascorbic acid sulphate sulphation of chondroitin, and finds it's very low. Perhaps it might be inorganic sulphate contaminant at that level.
Wilson: In your autoradiograph there were two spots in the brain. Can you tell us where they were? If they were in the pituitary gland, what was their significance?
Hornig: It is surprising (I didn't mention this before) that we have some radioactivity on the top of the tongue, but this also occurs if we give inorganic sulphate. We get almost the same appearance of radioactivity elsewhere which might suggest that this is derived from inorganic sulphate coming from ascorbic acid.
Wilson: Are you sure it was sulphate on the tongue and not ascorbic acid sulphate?
Hornig: We are not sure whether it was ascorbic acid sulphate *per se* incorporated as a molecule or whether it was just a sulphate moiety after splitting the bond between ascorbic acid and the sulphate and then taking the sulphate moiety to be incorporated somewhere else. We don't know. There is so little radioactivity we can locate in there.
Wilson: What was the significance of ascorbic acid sulphate in the pituitary gland, the tongue, and the intervertebral discs?
Hornig: In the pituitary gland and these other places, we find only very minute amounts of ascorbic acid sulphate.
Wilson: Still, it probably has some significance, hasn't it? Clinical information suggests that it may be ascorbic acid sulphate both on the tongue and on the intervertebral discs.
Brin: Our experience is that while unlabelled ascorbic sulphate is very highly stable, the labelled material, whether it is labelled in carbon or in sulphur, is rapidly decomposable; the incorporation of ^{14}C or ^{35}S results in breakdown, releasing inorganic sulphate to about 2–4% of total. Sulphation of chondroitin is minimal.
Hornig: We couldn't find that.
Lewin: Have you done any work on the phosphate of ascorbic acid?
Hornig: We have not done any work ourselves, but in the literature it is reported that ascorbic acid phosphate is a biologically 100% active derivative of ascorbic acid.

We don't have any experience with the phosphate. We have tried to synthesise ascorbic acid phosphate but it hasn't worked.

Degkwitz: Did you substitute the sulphate for the more normal vitamin C and feed it to guinea-pigs?

Hornig: To rats, yes, but not to guinea-pigs.

Degkwitz: Did they survive?

Hornig: No, they died. We gave them not more than 200 mg per day. If you give it intravenously the animals won't survive. The ideal dosage is guessed to be 50 mg per day. If you inject 200 mg the animal will be dead within the next few hours (this is in rats).

Degkwitz: You never looked at the cytochromes, did you?

Hornig: No.

Chairman: In the normal metabolism of ascorbic acid, it would appear from the very low levels of dehydro-ascorbic acid found that dehydro-ascorbic acid is very rapidly further metabolised once it is formed. Is that a correct deduction?

Hornig: Other workers have published work long before us, in the fifties, showing that ^{14}C-labelled ascorbic acid is oxidised to CO_2, and if you give ^{14}C-labelled dehydro-ascorbic acid, you will get, in a much shorter period, much greater amounts of ^{14}C-carbon dioxide.

Kotze: Since it is known that vitamin C is metabolised in guinea-pigs and man differently, don't you think that the function of vitamin C-2 sulphate between guinea-pigs and primates might also be different?

Hornig: This is what I mentioned. I said there is nothing known on the ascorbic acid sulphate metabolism—what happens with the sulphate—in man and monkeys. It is, for instance, not known whether ascorbic acid sulphate has any biological vitamin C activity in monkeys or man, as happens in trout. We can feed trout—salmon trout—on ascorbic acid sulphate, and it is biologically active without any ascorbic acid.

8

Vitamin C in Canning and Freezing

J. D. HENSHALL

The Campden Food Preservation Research Association, Chipping Campden, Gloucestershire, England

ABSTRACT

The role of ascorbic acid in the canning and freezing industries is discussed. The nature of the losses incurred on processing and storage of material is reviewed with special reference to fruit and vegetable products. It is shown that each product presents its own problems and that the retention of vitamin C activity is product-dependent. The use of ascorbic acid as a processing aid is reviewed briefly.

INTRODUCTION

Fruits and vegetables, either processed or 'fresh', are the source of dietary vitamin C for primates, and therefore a great deal of attention has been paid by food technologists to the ascorbic and dehydro-ascorbic acid contents of their products. Because of its multifunctional ability ascorbic acid has been used by food processors in the following ways, as:

(*a*) a vitamin;[1]
(*b*) an acid;[2]
(*c*) a curing aid;[3]
(*d*) an antioxidant;[3]
(*e*) an oxygen scavenger;[3]
(*f*) a colour stabiliser;[3]
(*g*) a bread improver;[3]
(*h*) a clarity improver in beer.[4]

In many of these applications products other than fruits and vegetables are involved, and therefore to some extent the range of sources of dietary ascorbic acid is extended. The obvious examples of this are where ascorbic acid is added in the canning and freezing of meat and fish.[3] On a practical basis in the United Kingdom, these uses are relatively minor, and can be dealt with in a few words. It has been known for thirty years that fish treated with a solution of ascorbic acid prior to freezing will have a better storage performance with respect to oxidative rancidity than untreated fish. The use of thickeners, such as carboxymethyl cellulose, enhances the effect of the ascorbic acid treatments.[3] However, as far as today's fish freezing and canning industry is concerned, this practice is not deemed to be necessary and is not carried out in the United Kingdom.

In the canning of meat and meat products ascorbic acid is used for packs such as large hams and chopped pork where a pasteurisation is given. The reasons for the addition of ascorbic acid are the same as for those products destined for shorter shelf-life packs and will be dealt with in the next paper. We can therefore confine ourselves to the study of ascorbic acid in relation to fruit and vegetable processing.

This study can be immediately divided into two sections:

(*a*) Where one can assess the retention, or otherwise, of vitamin C in the processing of a material.
(*b*) Where ascorbic acid is added as a processing aid.

There is a third category to be considered, and that is where fortification practices for nutritional purposes are involved, but I do not think that that topic falls within my brief.

NATURALLY OCCURRING VITAMIN C LEVELS

Let us consider the first aspect, *i.e.* the retention or otherwise of ascorbic acid in the processing of a fruit or vegetable. We have heard from Dr Kläui of the technical uses of added ascorbic acid and so we can see that it is in the interests of the processor to retain what ascorbic acid is already present in the raw material. It must be recognised from the outset, that if processing is going to be carried out, then there will be a loss of ascorbic acid, just as other food components will be lost. What I would like to do is to discuss how the ascorbic acid may be lost, how these losses may be minimised, and to describe the ranges of ascorbic acid that one may expect to encounter in the raw material. Taking the last point first, the concentration of ascorbic acid encountered in plant material is summarised in Tables 1 and 2.

The figures quoted are either mean values or the range found is given. The ascorbic acid contents will vary according to the cultivar, season, climate, maturity, duration of storage, etc. It should also be noted that the concentration of the vitamin will vary in different tissues of the plant.[7] Attempts have been made to increase the vitamin content of tomatoes by the selection of new cultivars, and a strong indication that this is possible has been made.[5]

LOSSES DUE TO PROCESSING

Reverting to the first point, *i.e.* how ascorbic acid may be lost, this has recently been described with special reference to canning.[8,9] Losses of ascorbic acid commence as soon as the fruit or vegetable is harvested. These losses are mainly due to enzymatic reactions of which peroxidase, ascorbic acid oxidase, cytochrome oxidase and phenolase are the most significant. The first three have been identified in citrus fruits.[10,11] In the tree-borne fruit it is supposed that these enzymes are balanced by reductase systems, which therefore control the level of ascorbic acid present at any one time. When the fruit is damaged and cellular fragmentation takes place reductases are more labile, and therefore the oxidases are free to react with the ascorbic acid.

Non-enzymatic changes which are of significance are the catalytic effects of copper, which are enhanced by iron, resulting in the formation of dehydro-ascorbic acid and hydrogen peroxide. The hydrogen peroxide produced in this reaction further reacts with ascorbic acid and the copper catalyst to give, directly or indirectly, oxygen and water.[12]

Dehydro-ascorbic acid has full vitamin C activity, but is more thermo-labile than ascorbic acid, and it is important that oxygen should be excluded as far as possible during processing. Furthermore, it is preferable that the processing of citrus juices is carried out in stainless steel apparatus, and any holding operation should be carried out in glass-lined equipment, as the main flavonoids of citrus juices do not possess the 3-hydroxyl-4-carbonyl group in the pyrone ring or the 3′,4′-dihydroxy group in the B-ring which is necessary for these compounds to complex with metal ions.

TABLE 1

Average ascorbic acid levels occurring in some fruits of interest to food processors[5]

Fruit	Ascorbic acid (mg/100 g edible portion)
Apple	2–10
Apricot	7–10
Blackberry	15
Blackcurrant	210
Damson	3
Gooseberry	40
Grapefruit	40
Lemon	50
Orange	50
Peach	7
Pineapple	25
Raspberry	25
Strawberry	60
Tomato	25

TABLE 2

Average ascorbic acid levels occurring in some vegetables of interest to food processors[6]

Vegetable	Ascorbic acid (mg/100 g edible portion)
Asparagus	25–33
Beetroot	5–12
Broad beans	25–30
Brussels sprouts	85–120
Carrots	6–11
Celery	5–15
Corn	10–15
Green beans	17–25
Peas	20–30
Potatoes	20[a]
Spinach	40–60

[a] Average over 12 months.

The behaviour of naturally occurring ascorbic acid in canned products has provided many research workers with a topic for investigation. I believe that there is scarcely a processed fruit or vegetable product which has not been examined at one time or another and a detailed analysis of all results is an impossible task because of the large number of variables involved in the experiments. However, a number of simple principles emerge from a critical look at some of the work. As mentioned earlier, there is an overall loss of ascorbic acid during the process, some of the unit processes contributing to a greater or lesser extent. Of these, the blanching of vegetables has received the most attention, and here the process details given in various papers are important. An overall view is that high temperature-short time (HTST) blanching gives the best retention of ascorbic acid. Varoquaux[13] in his work on 'medium' to 'extra small' petit pois showed, using precisely specified blanching conditions on a laboratory scale, that 97% ascorbic acid was retained after a 60 second blanch at 100°C. This decreased to about 88% retention if blanching was carried out for 4 min. The product: water ratio was given as 1:10. The usual commercial practice in pea blanching is to employ a 3–4 min blanch at temperatures of 90–95°C with a product: water ratio of 1:4 or 5. This work, while interesting, is of limited value when making an assessment of industrial processes. In a freezing or canning line peas are blanched continuously, and the rate of change of water is slow compared with the rate of throughput of product. That is to say, the peas are blanched in a solution which is more nearly isotonic than water, and there is little reliable published data on this point. Even though blanching can result in a severe loss of ascorbic acid, that which remains must be stabilised to some extent because air has been driven out of the tissues and any oxidative enzymes have been at least partially inactivated. Recent work on what is termed 'enzyme regeneration' has shown that an apparently inactivated peroxidase preparation can regain 10 and 30% of its original activity within 24 hr when stored at 1°C and 21°C respectively.[14] Peroxidase can lead to the oxidation of ascorbic acid,[9] and therefore this somewhat loosely termed phenomenon must be taken into account when assessing HTST processes, especially in the freezing industry.

The role of residual oxygen, arising from the product or headspace, is of importance in storage performance.[15] It was shown as long ago as 1945[16] that ascorbic acid, in cans of deaerated sweetened orange juice, was lost when the headspace contained oxygen. Deaeration of canned orange juice is now invariably practised, and the efforts of processors are directed towards the maintenance of a small headspace with a minimum amount of oxygen in the juice. It is not sufficient to fill orange juice hot and then seam, as hot juice may contain more oxygen than is present in the headspace of a properly filled and seamed can.

The effect of the container is of the utmost significance when discussing ascorbic acid retention with relatively high residual oxygen levels. Under conditions of low oxygen tension, it appears that the type of container is not really significant. In acid packs in plain cans oxygen will react preferentially with tin,[17–19] and thus greater amounts of ascorbic acid are retained in plain cans than in lacquered cans or glass jars after a similar period of storage. It was shown that dehydro-ascorbic acid values were only 1–2% of the total ascorbic acid, thus the ascorbic acid was breaking down, probably to diketo-gulonic acid. The effect of pH in the canned product is relatively minor as ascorbic acid is stable in the pH range encountered in canning.

Anaerobic destruction of ascorbic acid, following oxidative changes, is also significant.[20] The rate of this reaction is virtually independent of pH, except in the range 3–4 where it is slightly increased. Accelerators of this reaction are fructose, fructose-6-phosphate, and fructose-1,6-diphosphate, sucrose (probably following its hydrolysis to fructose), and caramelised fructose. Furfural and carbon dioxide appear to be the major products of decomposition. The more concentrated the fructose and related compounds, the quicker appears to be the loss of vitamin C.

TABLE 3

Percentage retention of ascorbic acid in canned orange juice at different concentrations.[21]

Storage temperature (°C)	Period (months)	Soluble solids (%)			
		13[a]	30	49	71
4·5	12	97	97	97	93
15·5	12	85	92	88	83
26·5	12	70	51	16	6
38·0	6	67	39	10	8
49·0	1	56	15	9	8

[a] Single strength juice contained 13% soluble solids.

It appears, in canned citrus products, that very little breakdown of vitamin C occurs during the heating process and this is also found to be so with other fruits. Significant losses begin to occur during storage. The rate of loss is time and temperature dependent and the results indicate that for optimum vitamin C retention, storage at 10°C or lower is preferred. One other factor besides oxygen, fructose and storage temperature is known to be significant, and that is that ascorbic acid is concerned in anthocyanin degradation. It has been shown that in canned anthocyanin-rich fruits, for example strawberries, 40–60% of the original vitamin C can be lost on processing. After 15 weeks storage at 37°C only 40% of the ascorbic acid content after processing remains. Saburov and Ulyanova[22] have reported that ascorbic acid may contribute to the polymerisation of anthocyanins in other products.

For canned vegetables, which may in some cases contribute a significant portion of the total dietary vitamin C, the same considerations apply, *i.e.* poor methods of preparation, high residual oxygen in the container, overprocessing, and subsequent storage at high temperatures will all contribute to the loss of vitamin C.

In the freezing industry the problems associated with attempting to retain the maximum amount of vitamin C are no less daunting than those facing the canner. Indeed in some ways they may be more so, as we shall see.

The losses which may be incurred by a freezer when processing fruits and vegetables are somewhat similar to those incurred by the canner as many of the unit processes are the same. Any major differences with regard to the final product can usually be ascribed to variations in plant layout and efficiency. The process of freezing in itself does not cause a loss of vitamin C. Evidence for this can be found in many published papers, *e.g.* Garry and Owen.[23] Once ascorbic acid is in a stabilising medium, as in 3% metaphosphoric acid solution, samples can be stored frozen prior to analysis. Without a stabilising medium problems of retention are manifested during storage, and it has been shown that handling during this stage is extremely important in this respect. Tables 4 and 5 illustrate this point.

It is important to realise that each individual product has its own characteristics with regard to ascorbic acid retention. As a sideline, the work by Hucker and Clarke[24] showed that blanched products could be held at relatively high temperatures for several hours prior

TABLE 4

Effect of thawing frozen vegetables on ascorbic acid content.[24]

Product	Thawing temperature (°C)	mg/100 g of ascorbic acid after (hr) holding				
		0	2	5	24	48
Corn	21·1	4·95	3·97	3·65	2·99	2·01
	7·2	5·80	4·45	3·75	2·66	1·85
	1·6	5·34	5·04	4·82	4·62	3·62
Green beans	21·1	10·11	8·13	8·12	6·72	—
	7·2	10·29	8·92	8·69	6·43	—
	1·6	8·99	7·93	7·38	6·00	—
Peas	21·1	18·99	17·93	16·79	14·19	12·45
	7·2	17·60	17·77	18·17	16·25	13·51
	1·6	17·45	18·84	18·15	16·90	16·38
Lima beans	21·1	21·10	20·19	21·43	15·34	12·96
	7·2	22·20	19·35	20·10	20·61	17·19
	1·6	21·40	21·00	20·61	20·53	18·61

to freezing without significant loss of ascorbic acid (*see* Table 6). Needless to say, this practice is condemned on bacteriological grounds, but the stabilising effect of the blanching process is illustrated.

Tables 4 and 5 emphasise the fact that the level of vitamin C in the product when received by the consumer can be controlled to a great extent by the retailer. He has the responsibility of ensuring that his storage and display cabinets are used in accordance with practices recommended by manufacturers. The major suppliers of frozen foods in this country ensure that their representatives are fully trained and are able to advise retailers as to the proper use of their cabinets.

We can therefore say that, given good processing conditions, the major loss of vitamin C in frozen foods occurs during storage. While this does not absolve the manufacturer from responsibility, it does mean that a hitherto largely unconsidered factor must be taken into account.

Even though only isolated examples have been cited from the large volume of literature on the subject, it is obvious that there is a wide range of factors to be taken into account when considering the natural vitamin C content of processed foods for human consumption. One final example of this is some data we have obtained in our laboratory, given in Table 7. These figures were obtained from a single commercial packet of frozen peas.

TABLE 5

The effect of alternate freezing and thawing on the ascorbic acid content of frozen vegetables.[24]

Number of freeze–thaw cycles	Thawing phase		mg/100 g of ascorbic acid			
	Time (hr)	Temperature (°C)	Corn	Green beans	Peas	Lima beans
0	0	21·1	5·46	7·82	16·97	21·10
1	1	21·1	5·38	6.84	16·02	20·00
2	2	21·1	3·59	7·82	15·64	21·10
3	1	21·1	3·02	7·06	15·00	20·00
4	2	21·1	2·75	6·32	15·13	19·30
5	1	21·1	2·47	6·75	15·07	20·50
6	3	21·1	2·09	6·95	15·03	19·10
0	0	7·2	5·13	6·33	17·28	22·02
1	45	7·2	2·57	6·96	16·62	21·90
2	21	7·2	2·38	6·35	15·81	19·30
3	52	7·2	1·86	6·06	12·11	18·30
4	19	7·2	1·15	5·28	11·29	17·10
5	7	7·2	0·92	5·66	10·85	15·90
6	5	7·2	1·01	5·98	10·00	16·00
0	0		5·39	7·61	17·89	21·40
1	141	1·6	2·49	6·43	12·76	18·40
2	108	1·6	1·05	5·33	8·71	12·70
5	44	1·6	1·50	4·07	2.86	6·60

TABLE 6

Effect of holding blanched vegetables prior to freezing on ascorbic acid content.[24]

Product	Temperature held prior to freezing °C	mg/100 g ascorbic acid after freezing — Hours held prior to freezing						
		0	2	4	8	12	24	8
Lima beans	21·1	19·04	18·79	17·79	16·67	17·87	18·24	—
	7·2	22·26	22·60	20·78	21·14	19·31	20·19	—
	1·6	19·04	20·79	18·79	19·39	19·69	19·31	—
Corn	21·1	4·98	4·54	3·75	3·75	3·75	—	—
	7·2	5·19	5·10	4·87	5·04	4·88	—	—
	1·6	5·64	4·65	5·44	4·74	4·80	—	—
Corn	21·1	14·75	16·75	14·80	14·82	11·77	12·25	9·06
	1·6	18·83	19·06	18·85	18·76	18·76	17·52	18·52

TABLE 7

The total vitamin C content of a sample of peas and the effect of variations in analytical procedure (mg/100 g)

	Indophenol assay		Fluorimetric assay
	Manual	Automated	Automated
Minimum	8·6	14·1	16·6
Maximum	15·4	15·6	19·4
Mean	11·2	14·7	18·0

Overall	Minimum	8·6
	Maximum	19·4
	Mean	14·9

ASCORBIC ACID AS A PROCESSING AID

We now come to the use of ascorbic acid as a processing aid in the canning and freezing industries. The subject has recently been reviewed[3] but an outline of the principal practices involved is necessary.

Certain fruits and vegetables become discoloured very quickly when disruption or injury of the tissue occurs. It has been shown that those tissues which discolour easily have a combination of low ascorbic acid levels and highly active phenolases.[25] At the cut surface a complex mixture of the phenolases and their substrates is formed and, in the presence of air, oxidation of the substrates takes place to orthoquinones. Although the phenolic structures of the various plant tissues vary, the dominant forms are chlorogenic acid, catechin, epicatechin, and leucoanthocyanins. The role of these compounds in various enzyme systems also varies, some of the phenolic compounds are substrates, some may be inhibitors, and some are inactive. The enzymes have multiple forms and have individual characteristics, for example those of apple, potato and mushroom.[26] However, as long as ascorbic acid is present, no browning occurs through the oxidation of the orthophenols to orthoquinones. Dehydro-ascorbic acid is ineffective in preventing this reaction. Dealing with the addition of ascorbic acid to canned fruit in order to produce a good quality product having a satisfactory shelf-life, there are documented cases of success, for example the well known work of Hope.[27] In this case the addition of 300 mg of ascorbic acid per pound of apple halves, enabled canning without an exhaust, controlled browning, reduced headspace oxygen, protected the container from corrosion, and increased the residual ascorbic acid content.

Bauernfiend and Pinkert[3] cite instances of other workers reporting that canned figs, apples, pears, and grapes all benefited from the addition of ascorbic acid to the syrup.

Some workers have reported different results with the addition of ascorbic acid, and its value would seem to be debatable. There is no doubt, however, that in the short term an improved product usually results, although ascorbic acid cannot be used to improve the quality of a product in which inferior grades of raw materials have been incorporated. A point of interest with regard to the addition of ascorbic acid to citrus juices is the apparent lower stability of the added vitamin compared to that which is naturally present.[28,29] Another worker has also alleged that increased corrosion of a metal container may result.[30] Both of these effects require further investigation.

Although the main traditional role of fruit has been to supply dietary ascorbic acid, the level present in most popular fruits such as peaches, apricots, pears, plums, nectarines, bananas and apples is relatively low when compared with citrus fruit. These fruits when frozen as slices or pitted generally discolour on thawing. A degree of off-flavour also results. This is most noticeable when the fruit has an open structure. (Reeve[31] has stated that 25% of the fruit volume of an apple is made up of air spaces.) A great deal of work has been carried out on this problem, for example inhibition or inactivation of the enzyme, the introduction of reducing agents, and the removal of air. The only additive which has a permanent effect is sulphur dioxide when used at a relatively high concentration. Ascorbic acid appears to be the most effective additive when applied by a vacuum treatment, but the concentrations must be high to be effective. In some cases ascorbic acid plus other reducing agents, such as cysteine,[32] appear to give a better effect, and again careful storage at controlled temperatures is a necessity for maintaining the effect.

In terms of tonnage sold, processed vegetables are now a significant part of the nation's diet. It has been reported that enzymatic browning in potatoes has been inhibited by ascorbic acid, and a 10–14 days shelf-life at 4°C has been obtained for the prepeeled product. The difficulty with potatoes is that different cultivars show a different response and that the same cultivar is not consistent in its performance. Ascorbic acid treatment of potato chips prior to partial cooking in oil before freezing has been shown to prevent undesirable darkening in the final product.[33] A point of interest with regard to harvesting is illustrated by reference to mechanically harvested broad beans, which sometimes suffer from bruising. The practice at the present time is to cool the beans in the field prior to transportation to the factory in order to prevent brown or grey discoloration caused by phenolase. Vacuum treatment with ascorbic acid will not prevent discoloration in this instance[34] and refrigeration is the only practical solution. Removal of oxygen leads to the growth of anaerobic bacteria and the development of off-flavours.

Canned vegetables are in general not affected by oxidative browning and there are few examples of changes caused by this effect. The best example here is the canned mushroom, the reducing action of the tin being insufficient to obtain the required light colour demanded by various quality standards, and thus ascorbic acid is added. A superior flavour is also claimed. In the case of canned cauliflower Chandler[35] recommends, other than suitable raw material, the use of lacquered cans, the avoidance of excessive heat treatment, and an ascorbic acid and sulphur dioxide treatment. Other packs reported to have benefited are canned carrots (prevention of carotene oxidation) and canned beets.[3]

CONCLUSIONS

To complete my brief survey of ascorbic acid in relation to canning and freezing, let us now try to assess the overall effect of processing. Should fortification be practised then there is no problem, but we must still answer the question, 'Is a processed pea or a helping of frozen spinach as good as "fresh"?'. In a recent article, Daniels[36] compared his analytical values for the ascorbic acid contents of garden fresh, market fresh, frozen and canned vegetables with those published in USDA Handbook No. 8,[37] and showed that the values quoted were useful only as a guide. Daniels' results are shown in Fig. 1.

This illustrates the danger of using published figures to compare processed and fresh foods when not all the factors involved are clearly defined. In particular it is important to distinguish between market-fresh and garden-fresh. The danger of using a standard text as a basis on which to formulate a diet are also highlighted. The

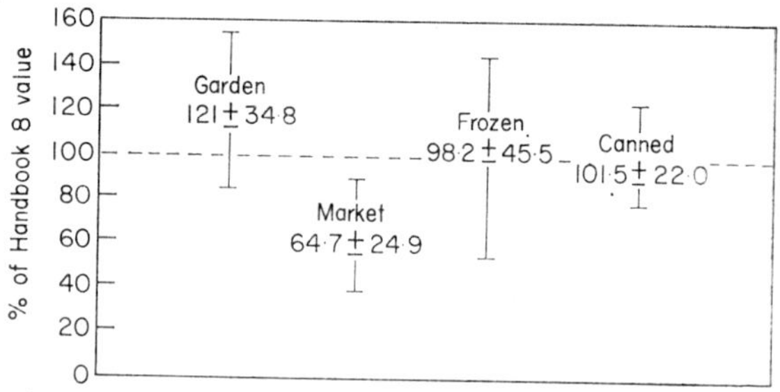

FIG. 1. Comparison of ascorbic acid content of garden fresh, market-fresh, frozen and canned vegetables with the corresponding values listed in USDA Handbook No. 8.[37]

values in a published list should only be used as an approximate guide. I have stressed that storage is of prime importance in the preservation of vitamin C and it must be remembered that so-called fresh vegetables have normally undergone some storage before reaching the consumer and possibly more before actual consumption. The ranges of vitamin C levels found are very large. Accepting this it is still possible to generalise and to say that, while garden fresh produce is the ideal, in the case of vegetables at least, canned and frozen products can be as good or better than market-fresh produce.

ACKNOWLEDGEMENT

The author is grateful to the Director of the Campden Food Preservation Research Association for permission to publish this paper.

REFERENCES

1. Anon. (1972). *Brit. Nutrition Foundation Bulletin*, No. 7. p. 8.
2. Clarke, K. J. (1970). *The Flavour Industry*, **1**(6), p. 388.
3. Bauernfiend, J. C. and Pinkert, D. M. (1970). *Adv. Fd Res.*, **18,** p. 219.
4. Urion, E., Chapon, L., Chapon, S. and Metche, M. (1956). *Am. Brewer*, **89**(12), p. 56 and p. 71.
5. Mapson, L. W. (1970). In *The Biochemistry of Fruits and their Products*, Vol. 1, (Ed. A. C. Hulme), Chapter 13, Academic Press, New York, p. 369.
6. Author's results.
7. Zilva, S., Kidd, F. and West, C. (1935). *New Phyt.*, **37,** p. 345.

8. Adams, J. B. and Blundstone, H. A. W. (1970). In *The Biochemistry of Fruits and their Products*, Vol. 2 (Ed. A. C. Hulme), Chapter 15, Academic Press, New York, p. 507.
9. Blundstone, H. A. W., Woodman, J. S. and Adams, J. B. (1970). In *The Biochemistry of Fruits and their Products*, Vol. 2 (Ed. A. C. Hulme), Chapter 15, Academic Press, New York, p. 543.
10. Vines, H. M. and Oberbacer, M. F. (1963). *Pl. Physiol., Lancaster*, **38,** p. 333.
11. Dawson, C. R. (1966). In *The Biochemistry of Copper* (Ed. J. Persach, P. Aisen and W. E. Blundberg), Academic Press, New York.
12. Weissberger, A. and LuValle, J. E. (1944). *J. Amer. Chem. Soc.*, **66,** p. 700.
13. Varoquaux, P. (1971). *Compt. Rend. Hebdom. des Seances de l'Acadamie d'Agriculture de France*, **57**(11), p. 945.
14. Wang, S. S. and Dimarco, G. R. (1972). *J. Fd Sci.*, **37,** p. 574.
15. Tressler, D. K. and Joslyn, M. A. (1954). In *The Chemistry and Technology of Fruit and Vegetable Juice Production*, New York, p. 175.
16. Boyd, J. M. and Peterson, G. T. (1945). *Ind. Engng Chem.*, **37,** p. 370.
17. Shiga, I. and Kimura, K. (1956). *Rpt. Oriental Canning Inst. Japan*, **4,** p. 55.
18. Khan, S. A., Ahmad, M. and Khan, S. (1963). *West Pakistan J. Agric. Res.*, **1**(3), p. 27.
19. Moore, E. L., Atkins, C. D., Huggart, R. L. and McDowell, L. G. (1951). *Citrus Ind.*, **32**(4), p. 5; **32**(5), pp. 8, 11, 14.
20. Kefford, J. F., McKenzie, H. A. and Thompson, P. C. O. (1959). *J. Sci. Fd Agric.*, **10,** p. 51.
21. Curl, A. L. (1947). *Canner*, **105**(13), pp. 14, 18.
22. Saburov, N. V. and Ulyanova, D. A. (1967). *Dakl. TSKHA.*, **132,** p. 259.
23. Garry, P. J. and Owen, G. M. (1967). *Technicon Symposium 'Automation' in Analytical Chemistry,* p. 507.
24. Hucker, G. J. and Clarke, C. (1961). *Food Tech.*, **15,** p. 50.
25. Diemair, W. and Postel, W. (1965). *Wiss. Veroeffentl. Deut. Ges. Ernahrung*, **14,** p. 248.
26. Cording, J., Eskew, R. K., Salinard, G. J. and Sullivan, J. F., (1961). *Food Technol.*, **15,** p. 279.
27. Hope, G. W. (1961). *Food Technol.*, **15,** p. 548.
28. Inagaki, C. (1943). *J. Agric. Chem. Soc. Japan.*, **19,** pp. 451, 464.
29. Matsui, O. and Ito, S. (1953). *J. Utilisation, Agric. Prod.*, **1,** p. 4.
30. Suzuki, K. (1971). *Canners Journal* (*Kanzume Jiho*), **50**(1), p. 104.
31. Reeve, R. M. (1953). *Fd Res.*, **18,** p. 604.
32. Walker, J. R. L. and Reddish, C. E. S. (1964). *J. Sci. Fd Agric.,* **15,** p. 902.
33. Hawkins, W. W., Leonard, V. G. and Armstrong, J. E. (1961). *Food Technol.*, **15,** p. 410.
34. Author's results.
35. Chandler, B. V. (1964). *J. Sci. Fd Agric.*, **16,** p. 11.
36. Daniels, R. W. (1974). *Food Tech.*, **28**(1), pp. 46, 60.

37. Watt, B. K. and Merrill, A. L. (1963). *Composition of Foods—Raw, Processed, Prepared*, rev. ed. Agriculture Handbook No. 8, U.S. Dept. of Agriculture, Washington, D.C.

DISCUSSION

Goodman: One brand of apple juice is described as having 'added vitamin C'. Is that an indication that the vitamin is nutritionally available to the consumer, or is it only a technological additive?
Henshall: That's a leading question! I think, to be quite honest, there is some element of truth in both those assumptions, that ascorbic acid will control the browning in apple juice and then, as a bonus, if you like, it is also nutritionally available; but it is there originally to control browning. Whether the level which is added is sufficient for that purpose or not I don't know.
Parducci: This question is directed towards the freezing of fruits and vegetables. Is there a temperature range over which the food processor should particularly avoid holding frozen vegetable materials during both freezing and thawing? More specifically, is there any evidence of enhancement of the rate at which frozen foods lose vitamin C when held at temperatures just below those at which ice begins to form?
Henshall: Yes—if I had had time to speak on the slides for longer, you would have had your answer, but there was just not time—there was too much data on the slides.
Kotze: In your introduction you mentioned that primates need ascorbic acid from an external source. It is known, however, that only higher primates need ascorbic acid, whereas the lower primates, like lemurs for example, synthesise their own. A paper in *Science* in 1973 indicated at which points during evolution certain species lost their ability to synthesise their own ascorbic acid.
Cooke: Is there much information to hand about vitamin C retention during the dehydration of fruits and vegetables?
Henshall: Not published, to any great extent. There are some figures for retention in some Government publications just after the last war, and there has been quite a lot of work done by individual firms. The problem here has been one of analysis, and for example with residual processing aids. Again it has been shown that the shorter the process, basically, the greater the retention of ascorbic acid.
Bointon: Can you specify the possible reason for the lower solubility of added ascorbic acid in fruit juices as opposed to natural ascorbic acid?
Henshall: One of the things I've been interested in is what in fact is the state of ascorbic acid within the plant, because everybody seems to assume that ascorbic acid is there, it's water-soluble, and so on. I was unable to find any particular reference to work on the actual state of combination, or how it is found within the plant. There was some work done about 1930, published in the *Biochemical Journal*, when people were first looking at

ascorbic acid oxidase, and some theories were put forward then, but since then I haven't been able to find anything very much. The role of the flavonoids has been mentioned, but again in this context I'm more concerned with citrus juices, and the citrus flavonoids, as I mentioned, do not have this ability of stabilising. I can find no mention of what state the ascorbic acid is in in the plant, and I should very much like to know. I'm also interested in ascorbic acid compounds—would these give any clues as to the state of combination?

Twomey: With reference to your last slide, was any consideration given to the different varieties of vegetables, and the season at which they were harvested?

Henshall: Yes. Daniels accumulated some data last year, so there was a gap of ten years between this and previous data. What has happened in that ten years is that varieties have changed, and not only have varieties changed but the processing industry has changed and the areas of production have changed, so that there was a longer haul for the market produce from the areas of production to the site of consumption. People are still quoting the ten-year old values whereas these values may not necessarily have been true.

Kläui: You showed that ascorbic acid was less stable in orange juices with a high soluble solids content. Do you think that this is due to only the presence of sugars (and other soluble solids), or might it also be connected with processing conditions, such as the process of concentration?

Henshall: In the work which I quoted there, they took one sample of the concentrate and then diluted it; they correlated the changes with the fructose-type compounds. The only thing that was investigated, in fact, was the fructose and its related compounds.

Rolfe: You mentioned that the flavonoids in citrus fruits do not exert an action on vitamin C as they do not chelate with copper. It is true that copper is a very potent catalyst for the oxidation of ascorbic acid, and except for the trace amounts that are always present in ordinary circumstances, the vitamin would be much more stable towards oxidation. Inactivation of copper chelation will therefore confer stability.

However, it is to be expected that the citric acid of citrus fruits will complex with any traces of metal catalyst present, and as a consequence the requirement that in order to exert protective action the flavonoid should complex with copper is not of prime importance. It is possible that the autoxidation of AA proceeds via a chain reaction, and the protective action of flavonoids follows a similar behaviour to that postulated for their protective action in autoxidation of fats. In the latter case, we observe the formation of hydroperoxides, and, in the case of ascorbic acid, hydrogen peroxide. The activity of flavonoids in the protection of vitamin C thus becomes one of a free radical acceptor and to break chain reactions. There is some published work in support of this theory relating to the protection by the flavonoids in blackcurrant juice of vitamin C.

Henshall: Yes, I was a little wary of that one when I made the statement. But if you look at the hypotheses which have been put forward, the disappointing fact is that nobody really seems to know what happens. This has come up in other papers, I believe. I think this type of work should be pursued because I think it is very interesting, and we would learn a lot from it from the point of view of increasing the efficiency of processing, because it is in this area that the process can be blamed—sometimes I think unfairly—for losing ascorbic acid, when in actual fact it could be a natural reaction.

Rolfe: There has been a small amount of work published on this subject. I did some myself at the University of Strathclyde, with blackcurrant juice.

9

The Significance of Ascorbates and Erythorbates in Meat Products

M. D. RANKEN

British Food Manufacturers Research Association, Randalls Road, Leatherhead, Surrey, England

ABSTRACT

The natural content of vitamin C in meat is very low. The application of ascorbic acid or ascorbates to fresh meat at concentrations of the order of 100 *ppm delays the development of brown colour. This use is prohibited in the UK but it is permitted in fresh meat products such as sausages or hamburgers. Ascorbates may also be used, in similar concentrations, to accelerate and intensify the action of nitrites in forming the red pigment in cured meats, nitrosyl myoglobin. The nitrite and ascorbate interact, giving lower concentrations of both in the final product. From certain points of view, notably avoidance of nitrosamines, low nitrite concentrations are considered advisable, but there are chemical and microbial penalties if it falls too low. There are currently strong suggestions that the use of very high levels of ascorbate, such as* 1000 *ppm, eliminate or greatly reduce nitrosamine formation in some meat products. This effect is due in part to reduction in the level of available nitrite, but the ascorbate may also inhibit the nitrosation reaction in meat. There may, however, be severe penalties,* e.g. *total discoloration of the product, in some circumstances.*

Erythorbate is currently cheaper than ascorbate and can be used for some of the purposes mentioned above. The balance of the evidence is that apart from its lack of vitamin activity its chemical behaviour is similar to that of ascorbate.

Ascorbyl palmitate is used as a fat antioxidant in meat products such as fresh sausage, at levels of about 0·5% *in the fat, and ascorbic acid may also be used for its antioxidant properties in certain specialised applications.*

The contribution of vitamin C from these dietary sources does not now appear to be negligible.

INTRODUCTION

The natural vitamin C content of fresh meat, with the exception of liver, kidney and brain which may contain 100–300 ppm, is usually considered to be negligible.[1,2] Russian workers give values up to

100 ppm for some apparently normal meats.[3] Ascorbic acid and related substances are not usually added to meat for nutritional reasons, and additions are only sometimes declared or advertised in such terms. Such treatment is always for technical reasons, almost always concerned with retention of the colour of the meat, and dates back to the discovery in America of its efficacy in 1949[4] and to the permission granted by the USDA in 1955 for these materials to be used, in specified amounts, in certain meat products.

The substances which are commonly used are ascorbic acid, erythorbic acid and their sodium salts. If one allows for the difference in pH, the acids and their salts behave similarly: ascorbates and erythorbates also have many similarities which will be considered more closely later. For convenience in this review, except where the context requires otherwise, the four substances will be considered together and referred to as 'ascorbates'.

The complex subject of the relationships between ascorbates and meat colour has been reviewed recently by authors in Germany,[5–7] Italy[8] and the USA.[9] The purpose of this paper is to summarise the main features, to note some very recent developments and to draw attention to certain aspects which are particularly relevant under British conditions.

FRESH MEAT COLOUR

The character of the red colour of fresh meat is mainly controlled by the balance between three forms of the meat pigment myoglobin, as depicted in the upper part of Fig. 1. At the surface of fresh meat, in contact with oxygen, the colour is the bright red of oxymyoglobin; in the anaerobic interior it is purple myoglobin: myoglobin is rapidly oxygenated and converted to oxymyoglobin when a freshly cut surface is exposed to air. Between the red and purple zones, where the oxygen tension is positive but small, the more stable brown metmyoglobin is formed. The metmyoglobin layer gradually increases in thickness, on the side nearer to the surface, as oxygen is consumed in the body of the meat. The red layer being transparent, the appearance of the surface therefore changes, over $\frac{1}{2}$–2 days, to the unappetising brown colour of old meat.[10,11] The effect of strong reducing agents, such as ascorbate, at low temperatures (0–5°C) and at moderate concentrations (*e.g.* 200 ppm) is to inhibit the formation of metmyoglobin and prolong commercial storage life by about 1 day.[12,13]

The oxidation of myoglobin to metmyoglobin is also catalysed by peroxides present in oxidising fat. Ascorbate by itself catalyses fat peroxidation[9,14] but the effect on colour caused by this is not normally great enough to offset the beneficial direct action. In the presence of added antioxidants, however, ascorbate acts synergistically to retard oxidation of both fat and pigments.

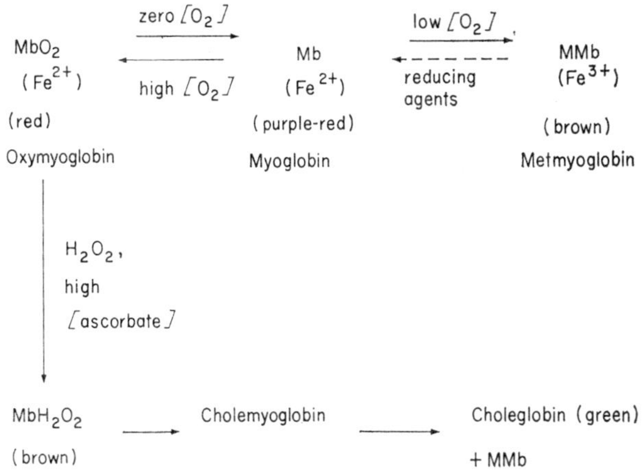

FIG. 1. Colour changes in fresh meat.

At concentrations over *ca.* 500 ppm, in air and especially at higher temperatures, ascorbate may form a dark coloured hydrogen peroxide–myoglobin complex which breaks down quite rapidly to a green pigment.[15,16] This effectively limits the concentration of ascorbate which might be used in practice to about 200 ppm.

In the UK, treatment of fresh butchers' meat with ascorbates is considered to be to the prejudice of the customer and is prohibited by the Meat (Treatment) Regulations, 1964, though Möhler[17] considered that the colour of ascorbate treated meat has a pinkness which would be an effective bar to deceptive practice. The use of ascorbic acid or ascorbates in meat products such as sausages or hamburgers is not prohibited in the UK but is probably not widespread. Of other countries, only Belgium, Holland, New Zealand and Sweden appear to permit the use of ascorbates for this purpose.[9]

CURED MEAT COLOUR

Role of Nitrite

The pigments of most importance here are nitrosyl myoglobin, NOMb, the deep red pigment found in raw bacon, raw ham and certain uncooked cured sausages, and nitrosyl myochrome $(NO)_2Mc$, which is the rather pinker colour of cooked cured meats such as frankfurters, cooked ham, luncheon meat or fried bacon. Though the route of formation of these pigments is still not certain, there is fair agreement that the main pathway is as shown in Fig. 2, at

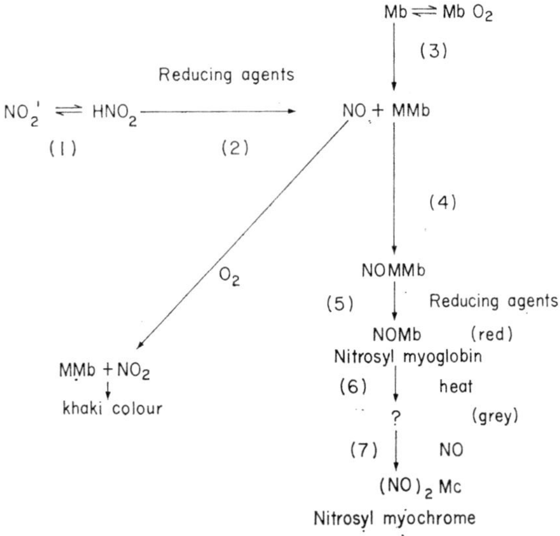

FIG. 2. Colour changes in cured meat. (The numbers indicate steps referred to in the text.)

any rate as far as Step 5 shown there.[6,18] Experience in our laboratory leads us to suggest the existence of the unnamed substance shown at Step 6, not previously discussed in the literature, for it can readily be observed when a piece of raw cured meat is slowly heated that the colour does not change directly from raw red to cooked pink but passes through an intermediate grey stage. At Step 7, nitrosyl myochrome is represented with two nitrosyl groups, in accord with the opinion of Tarladgis,[19] though the evidence for this is not certain.[18]

Frouin and Cordier[20] have recently presented evidence which implies that nitrosyl myochrome may contain only one nitrosyl group.

Role of Ascorbate

Ascorbate aids the development of the cured colours in a number of ways.

(i) Steps 2 and 5 in the reaction sequence (Fig. 2) are reduction reactions, normally effected by reducing systems present in the meat, such as sulphydryl compounds[21] or the cytochrome system.[22] It is to be expected that a strong reducing agent such as ascorbate will accelerate both steps.

(ii) Conversely, it is known that the presence of oxygen in sausage mixtures inhibits development of colour, and the oxygen scavenging action of ascorbate removes this disability.[23] In canned luncheon meat it has also been shown to diminish the brown colour formed as a result of oxidation of pigments to metmyoglobin.[24]

(iii) Ascorbate not only accelerates the formation of nitric oxide (Step 2) but by taking part in the reaction with nitrous acid it increases the net yield, permitting one to obtain a given degree of coloration with a saving of about one third in the quantity of nitrite required.[5] The intensity of the colour is improved where the nitrite level is low, but not where the concentration of nitrite would be adequate for full colour formation without the addition of ascorbate.[8,9]

(iv) It has been suggested that ascorbate may participate directly in the colour forming process, by forming a complex with nitric oxide which assists the formation of nitrosyl metmyoglobin at Step 4,[18] or by incorporation in the structure of nitrosyl myochrome at Step 7.[20]

For these reasons ascorbate is now widely used as a 'cure accelerator' for meat products such as cured cooked sausages of the frankfurter type, pasteurised ham, luncheon meat, etc.; 150–200 ppm of added ascorbate appears to be sufficient.[5]

In the presence of air and large excess of ascorbate, *e.g.* 1000 ppm ascorbate to 100 ppm nitrite, severe brown discoloration may occur in cooked or uncooked cured meat. Nitric oxide liberated from the nitrite becomes oxidised to nitrogen dioxide which forms a khaki-brown addition product with myoglobin. 'Nitrite burn' may be caused by a similar reaction when there is a considerable excess of nitrite, even in the absence of ascorbate.[25,26]

The presence of ascorbic acid also has a beneficial effect on the stability of the cured colour once it has beeen formed. Again, several factors are involved.

(i) Breakdown of the cured pigments, whether uncooked or cooked, appears to commence with dissociation of nitric oxide from the molecule.[27] One would predict, from mass action considerations, that this dissociation should be inhibited by the excess nitric oxide produced by the reaction between ascorbate and any residual nitrite.

(ii) Dissociation of the pigment is also favoured by the presence of fat peroxides:[27] as in fresh meat,[15] the formation of these is inhibited by tocopherols occurring naturally in the meat or by added antioxidants and ascorbate acts synergistically to improve the action of these.[9]

(iii) Colour stability is better in the absence of oxygen and is therefore improved by the oxygen scavenging action of ascorbate.

In vacuum packed cured meats 100–200 ppm of ascorbate is adequate for colour stability.[28] Higher concentrations further increase stability, but less than proportionally: levels up to 1000 ppm have been suggested.[29] Chelating agents such as EDTA[30] or citrate[31] increase the effect.

ASCORBATE AND NITROSAMINES

There is now evidence that the use of high concentrations of ascorbates, such as 1000 ppm or more, may be helpful in reducing the risk of nitrosamine formation in cured meats. Fiddler *et al.*[32] in the USA showed this with 550 and 5500 ppm ascorbate in frankfurters. They used nitrite concentrations well above the legal limit, together with drastic heating processes, to ensure nitrosamine formation in the control frankfurters. Fazio *et al.*,[33] also in the USA, demonstrated the production of nitrosamine in well cooked bacon and ascorbate at 1000–2000 ppm has been found effective in reducing the formation of nitrosamines in fried bacon.[34]

PROBLEMS WITH BACON

The American bacon used in the experiments just referred to differs from its British (or Danish or Polish) equivalent in a number of respects. First, Americans commonly fry bacon to a more crisp condition than is usual in Britain, and it is probable that nitrosamine formation is related to the temperature of cooking. Another difference is that a high proportion of American bacon, perhaps nearly 100%, receives a heat treatment by smoking or pasteurising, which has the effect of converting the cured pigment to the cooked form at the manufacturing stage. It is usual to vacuum pack the product. In this country, although some bacon is made by similar processes to this, the great majority receives no heat process until it is cooked for consumption. This means that the cured colour remains in the uncooked form throughout storage, distribution and retail sale and also during domestic storage after purchase. The majority of British bacon is not vacuum packed, and of that which is, a significant proportion is likely to be kept for up to a few days in the home after the pack has been first opened and some product removed.

The technological studies so far referred to were concerned almost entirely with cooked cured meats, and a number of complications arise, almost uniquely, in the case of bacon distributed and sold under UK conditions, in the uncooked state.

The colour of this bacon, and certain other questions relating to the levels of nitrite present, should be considered at each of several successive stages.

(a) Development of the Uncooked Cured Colour

As discussed above, the formation of nitrosyl myoglobin requires nitric oxide, normally supplied from added nitrite by the action of enzymes or reducing agents endogenous to the meat or added to accelerate or intensify the reaction. Added ascorbate increases the conversion of nitrite to nitric oxide and therefore gives a deeper red colour more quickly and uniformly, at the same time reducing the level of residual nitrite. The reaction does not always go to completion and residual nitrite and ascorbate may coexist in the product. Considerable excess of ascorbate may produce the unpleasant brown colours referred to earlier.

(b) **Stability of the Uncooked Cured Colour in the Absence of Air**

As already indicated, colour stability under these conditions is favoured by those factors which ensure the presence of nitric oxide in contact with the cured meat. The conversion of residual nitrite to nitric oxide by the action of ascorbate is thus beneficial in this way. On the other hand, the depletion of the residual nitrite concentration may diminish the microbial stability of the bacon since nitric oxide[35] and probably also ascorbate have no significant antimicrobial effect in these conditions.

(c) **Stability of the Uncooked Colour in Air**

This condition occurs at the surface of a block or joint of bacon or of unpacked slices exposed for sale or stored domestically. Such bacon when containing ascorbate at an initial concentration of 200 ppm has been observed in our laboratory to be decolorised in $\frac{1}{2}$–1 day at room temperature (*ca.* 15°C), even in the presence of 10–20 ppm residual nitrite. The mechanism of this change is not yet known.

(d) **Development of the Cooked Cured Colour**

It has been noted that the transition from uncooked colour to cooked colour takes place in at least two stages, the first of which is the loss of the original uncooked colour. It is clear from our experiments[36] that the last stage, the formation of pink nitrosyl myochrome from the intermediate grey substance, requires an adequate supply of nitric oxide at the time of cooking. Some nitric oxide may be provided from the original nitrosyl myoglobin but this is not normally sufficient without also a source from residual nitrite. Gaseous nitric oxide, produced by decomposition of residual nitrite in a vacuum pack, does not appear to be effective; perhaps because it is lost in the early stages of the cooking process. In the absence of ascorbate, a residual nitrite content of about 30 ppm *at the time of cooking* is necessary to give good uniform pink cooked colour, and the colour is quite unsatisfactory at levels below 15 ppm. To ensure 15 ppm after long but otherwise satisfactory storage under both trade and domestic conditions, the residual nitrite content at the time of manufacture may require to be as high as 80 ppm.

The presence of ascorbate in the bacon is detrimental because it diminishes the amount of nitrite available at the time of cooking, the nitric oxide formed from the nitrite lost not being of use. Although the stability of the uncooked nitrosyl myoglobin is improved under vacuum by the action of ascorbate so that a slightly increased supply of nitric oxide might be available from this source, the increase is not sufficient to offset the penalty caused by the reduction in residual nitrite content.

(e) Stability of the Cooked Cured Colour in the Absence of Air

Much of the literature on the stability of 'cured colour' refers to this situation, as for example in the interior of products such as ham. Pasteurised or hot-smoked prepacked bacon are similar in this regard, but ordinary cooked bacon is rarely stored under these conditions.

Ascorbate provides little benefit at higher levels of nitrite, *e.g.* 200 ppm, but is helpful when the nitrite content is low.[16,23,28] This suggests that the effective agent may again be nitric oxide.

The microbiological situation is complicated by the formation of new antimicrobial factors on heating meat in the presence of nitrite[37,38] and it is not known what additional influence the presence of ascorbate may exert here.

(f) Stability of the Cooked Cured Colour in Air

This also has been widely studied in ham, frankfurters, etc., and is applicable in some circumstances both to pasteurised and to cooked bacon. Fox *et al.*,[23] working with frankfurters, and Taylor[39] with chopped pork after canning, found that the presence of air was highly detrimental to the cooked pigment and that ascorbate improved the stability, presumably by scavenging oxygen. Rongey *et al.*[29] showed a colour stabilising effect of ascorbate in luncheon meats exposed to the air, even if containing no residual nitrite. A similar beneficial effect has been observed by us in cooked bacon.[35] The reason is not known, but it clearly cannot be an effect due to nitric oxide.

Summary

The several influences of ascorbate on bacon colour are summarised in Table 1.

TABLE 1

Summary of effects of added ascorbate on the colour of bacon

	Formation	Stability in absence of air	Stability in air
Uncooked colour	Improved rate and uniformity Improved intensity when nitrite inadequate (mainly due to NO) Destroyed by excess + air	Improved (due to NO + antioxidant)	Destroyed
Cooked colour	Diminished after storage in vacuum Destroyed after storage in air (mainly due to loss of NO_2)	Improved (due to NO + antioxidant)	Improved (? specific effect)

Other Problems

It is most important to use sodium ascorbate and not ascorbic acid when making up curing brines, otherwise at the low pH of ascorbic acid nitric oxide will be produced in the brine. On evolution into the air this is converted into highly toxic brown fumes of nitrogen dioxide.[5,9]

ASCORBYL PALMITATE

Ascorbic acid accelerates the oxidation of meat fats unless a tocopherol or metal chelating agent is also present.[9,14] Furthermore the use of either ascorbic acid or sodium ascorbate as an antioxidant in meat or meat products is limited by the practical difficulty of distributing these water soluble materials through the complex structure of animal fatty tissue.

Ascorbyl palmitate has some use as an antioxidant or synergist of other antioxidants, being more fat soluble than the ascorbates.[40]

The British sausage, containing a proportion of bread rusk, may suffer from a condition known as 'white spot' in which chalky white patches are formed on the surface, apparently as a result of oxidative change. Ascorbyl palmitate is a constituent of a proprietary mixture used in the UK, at a level of about 0·3 ppm in the sausage, to control this phenomenon.[41]

COMPARISON BETWEEN ASCORBATE AND ERYTHORBATE

Figure 3 shows the structures of the isomers of ascorbic acid.

Erythorbates are cheaper than ascorbates and in so far as their properties are similar there may be a good case for their use.

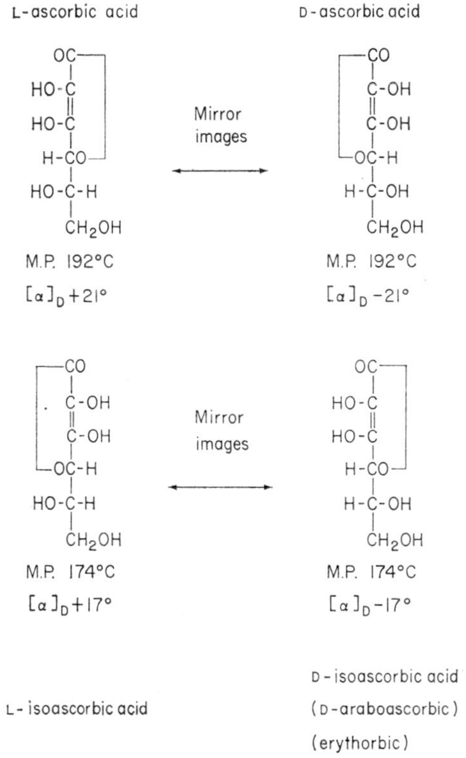

FIG. 3. Isomers of ascorbic acid (after Borenstein[42]).

Borenstein[42] and Trepow and List[40] reviewed the general behaviour of ascorbic and erythorbic acids and came to the same general conclusion, that ascorbic is a more efficient antioxidant than erythorbic. In model system experiments using meat it has been found that ascorbate was slightly more effective than erythorbate in accelerating the formation of cooked colour, but that their longer term effects on colour stability were similar.[43] In practical trials of various kinds, little or no difference has been found between the two isomers. Reith and Szakály[13] found no difference in a model system in which they measured fresh meat colour; Mills *et al.*[44] and Fox *et al.*[23] found no difference in effect on the formation and stability of (cooked) colour in frankfurters; Fiddler *et al.* found similar properties with respect to inhibition of nitrosamine formation in model systems (though here there was a slight advantage to erythorbate)[45] or in frankfurters.[32]

The antiscorbutic activity of erythorbic acid on the other hand is very low, only 1/20 or 1/40 that of ascorbic.[40] However, as there does not appear to be any reason to regard the amounts of ascorbate currently added to meat products as critical from the nutritional point of view, and in the absence of any evidence that moderate amounts of either ascorbate or erythorbate in the diet are harmful, there seems little reason to avoid the use of erythorbate on grounds only of its lack of vitamin activity. In the USA it is permitted as a complete alternative to ascorbate in meat products. In the UK it is not permitted as an antioxidant but is not otherwise prohibited. Bauernfiend and Pinkert state that it is prohibited in Austria, Belgium, Brazil, France, Germany, Holland, Italy, Norway, Sweden and Switzerland.[9]

Ascorbyl palmitate exhibits full vitamin activity in its ascorbyl portion, so that on a weight basis it has 39% the vitamin activity of ascorbic acid.

CONCLUSION

Two features of immediate interest emerge from this review. The first is that, whatever may be the advantages of using high proportions of ascorbate or ascorbic acid to minimise nitrosamine formation in or by other meat products, the consequences of such additions to bacon handled in the traditional British way include certain severe penalties to the quality of the product. Advantages to bacon quality include:

> improvements in the colour of uncooked vacuum packed bacon or, presumably, bacon in the anaerobic interior of a large piece such as a side, and in the anaerobic stability of that colour;
> improvement in colour formation and stability in pasteurised vacuum packed bacon and also in the stability of this colour on exposure to air.

The disadvantages, however, are the complete decoloration of uncooked bacon exposed to the air and the failure of some bacon, even if red before cooking, to form a pink colour when cooked after exposure to the air for more than a few hours. The circumstances of storage, etc., under which these disadvantages may be manifested are not universal but neither are they at all uncommon in current British practice; at present it seems that they may be avoided only by radical change in our present commercial and domestic ways of handling bacon.

Secondly, it is clear that the ascorbates added to meat products may already be significant contributors to the vitamin C content of the British diet, although there is only an insignificant amount in the meat itself.

A consumer of bacon or ham containing residual sodium ascorbate at the level of 100 ppm, purchasing the national average of about 6 oz per week,[46] would obtain 15 mg of ascorbic acid per week by this route or about 10% of his nutritional requirement.

Vitamin C deficiency is a feature of British diets only in exceptional cases, not likely to be influenced by its availability from meat products, so this source need not be regarded as nutritionally important. It is clear, however, that if there should be intense and widespread use of ascorbates, at concentrations of around 1000 ppm, in connection with the nitrosamine question, coupled with extensive revision of our handling methods so as to avoid the evident disadvantages, then the whole nutritional requirement of non-vegetarians might easily come from bacon alone.

REFERENCES

1. Davidson, S. and Passmore, R. (1966). *Human Nutrition and Dietetics*, 3rd edn, Livingstone, London.

2. McCance, R. A. and Widdowson, E. M. (1967). *The Composition of Foods*, H.M.S.O., London.
3. Danilov, M. M. (1969). *Handbook of Food Products: Meat and Meat Products*, Israel Program for Scientific Translations, Jerusalem.
4. Chang, I. and Watts, B. M. (1949). *Fd Technol.*, **3,** p. 152.
5. Grau, R. (1969). *Proc. Inst. Fd Sci. & Technol.*, **2,** p. 43.
6. Möhler, K. (1970). *Z. Lebensm. -Unters. u. -Forsch.*, **142,** p. 169.
7. Möhler, K. (1972). *Z. Lebensm. -Unters. u. -Forsch.*, **147,** p. 123.
8. Frati, G. (1972). *Industria Conserve*, **47,** p. 200.
9. Bauernfiend, J. C. and Pinkert, D. M. (1970). *Adv. Fd Res.*, **18,** p. 219.
10. Solberg, M. (1970). *Can. Inst. Fd Technol. J.*, **3**(2), p. 55.
11. Heiss, R. and Eichner, K. (1969). *Fleischwirtschaft*, **49,** p. 757.
12. Caldwell, H. M., Glidden, M. A., Kelley, G. G. and Mangel, M. (1960). *Fd Res.*, **25,** p. 139.
13. Reith, J. F. and Szakály, M. (1967). *J. Fd Sci.*, **32,** p. 188.
14. Lawrie, R. A. (1966). *Meat Science*, Pergamon Press, Oxford.
15. Greene, B. E., Hsin, I-M. and Zipser, M. W. (1971). *J. Fd Sci.*, **36,** p. 940.
16. Fox, J. B. (1966). *J. Ag. Fd Chem.*, **14,** p. 207.
17. Möhler, K. (1956). *Dt. Lebensm. -Rundsch.*, **52,** p. 179.
18. Fox, J. B. and Ackerman, S. A. (1968). *J. Fd Sci.*, **33,** p. 364.
19. Tarladgis, B. G. (1962). *J. Sci. Fd Agric.*, **13,** p. 485.
20. Frouin, A. and Cordier, J. P. (1973). *XIX Mtg Meat Res. Wkrs, Paris*, **4,** p. 1473.
21. Mirna, A. and Hofmann, K. (1969). *Fleischwirtschaft*, **49,** p. 1361.
22. Walters, C. L. and Taylor, A. McM. (1965). *Biochim. Biophys. Acta.*, **96,** p. 522.
23. Fox, J. B., Townsend, W. E., Ackerman, S. A. and Swift, C. E. (1967). *Fd Technol.*, **21,** p. 386.
24. Hardy, P. W., Blair, J. S. and Kreuger, G. J. (1957). *Fd Technol.*, **11,** p. 148.
25. Bacus, J. N. and Diebel, R. H. (1972). *Appl. Microbiol.*, **24,** p. 405.
26. Reith, J. F. and Szakály, M. (1967). *J. Fd Sci.*, **32,** p. 194.
27. Tarladgis, B. G. (1961). *J. Am. Oil Chem. Soc.*, **38,** p. 479.
28. Fredholm, H. (1967). *Fleischwirtschaft*, **47,** p. 1340.
29. Rongey, E. H., Kahlenberg, O. J. and Naumann, H. D. (1959). *Fd Technol.*, **13,** p. 640.
30. Borenstein, B. and Smith, E. G. (1968). U.S. Pat. 3,386,836.
31. Anon. (1971). *Fleischerei*, **22**(5), p. 42.
32. Fiddler, W., Pensabene, J. W., Piotrowski, E. G., Doerr, R. C. and Wasserman, A. E. (1973). *J. Fd Sci.*, **38,** p. 1084.
33. Fazio, T., White, R. H., Drusold, L. R. and Howard, J. W. (1973). *J.A.O.A.C.*, **56,** p. 919.
34. Herring, H. K. (1973). *XIX Mtg Meat Res. Wkrs, Paris*, **4,** p. 1517.
35. Shank, J. L., Silliker, J. H. and Harper, R. H. (1962). *Appl. Microbiol.*, **10,** p. 185.
36. Evans, G. G. (1974). *B.F.M.I.R.A. Res. Rept.*, in preparation.
37. Ashworth, J. and Spencer, R. (1972). *J. Fd Technol.*, **7,** p. 111.

38. Ashworth, J., Hargreaves, L. and Jarvis, B. (1973). *J. Fd Technol.*, **8,** p. 477.
39. Taylor, A. McM. (1961). *B.F.M.I.R.A. Res. Rept.*, No. 106.
40. Trepow, H. and List, D. (1972). *Ind. Obst. Gemusewert*, **57,** p. 29.
41. Evans, G. G. (1971). *B.F.M.I.R.A. Tech. Circ.*, No. 486.
42. Borenstein, B. (1965). *Fd Technol.*, **19,** p. 1719.
43. Kellam, G. (1966). *B.F.M.I.R.A. Tech. Circ.*, No. 321.
44. Mills, F., Ginsberg, D. S., Ginger, B., Weir, C. E. and Wilson, G. D. (1958). *Fd Technol.*, **12,** p. 311.
45. Fiddler, W., Pensabene, J. W., Kushnir, I. and Piotrowski, E. G. (1973). *J. Fd Sci.*, **38,** p. 714.
46. MAFF, *Household Food Consumption and Expenditure,* 1970–71, H.M.S.O., London.

DISCUSSION

Kläui: You mentioned that ascorbic acid may accelerate the formation of peroxides in fats. I think this is possible only if at the same time catalysts which act as oxidation-promoters are present, such as iron, copper and haemoglobin.

Ranken: Yes, certainly iron or metals are known to be involved in this, and if we put a chelating agent there it gets much better.

Degkwitz: It is commonly said that iso-ascorbic acid has no vitamin activity. But it really seems to be a problem of different pharmacokinetics, since it is very difficult to get iso-ascorbic acid into the organs. And it seems to be a sort of competition too, since it is far easier to get iso-ascorbic acid into deficient guinea-pigs.

Ranken: Dr. Hughes, in a Symposium here last year, read a paper on this topic, and he said I think what you're saying, that this was a question of transport of the molecule across membranes, and he put forward then a hypothesis that related the difficulty of getting iso-ascorbate into organs and across membranes to its steric shape as the difference from ascorbate. I am interested in what you say, that where there is deficiency it is likely to go in.

Hughes: Yes, that is what in essence we have found, that if you can get the iso-ascorbic acid into the tissues then it does have full vitamin C activity. The earlier reports of iso-AA having a biological potency only one quarter of that of AA are really a reflection of its different rates of gastro-intestinal absorption.

10

Vitamin C in Soft Drinks and Fruit Juices

D. M. GRESSWELL

Research and Development Department, Beecham Products, The Royal Forest Factory, Coleford, Gloucestershire, England

ABSTRACT

Citrus and other fruit juices have long been associated with vitamin C, and this has led to the development of soft drinks which have improved nutritional value by virtue of their vitamin C content. Such products generally contain sufficient fruit juice to support any label claim for vitamin C, so it is important that the processing of the juice should ensure the maximum retention of the naturally occurring vitamin C. Many factors determine the initial quantity and rate of degradation of the vitamin, ranging from fruit quality to trace metal contamination. It may be necessary to supplement the natural vitamin content by the addition of synthetic vitamin C, to make good processing losses, and provide for any further losses during distribution and in the consumer's home. L*-ascorbic acid (and iso-ascorbic acid) can be effectively used as an antioxidant for achieving improved flavour stability during processing or during the shelf-life of soft drinks and fruit juices. Under certain circumstances, the presence of vitamin C does have some undesirable effects such as the bleaching of natural and artificial colourants, non-enzymic browning reactions, etc. The nature of the product packaging can also affect the stability of the vitamin, or make its role as an antioxidant very important. In any application, analytical quality control of vitamin C content is required for practical, economic and legal purposes. The legislation controlling the use of vitamin C in soft drinks and fruit juices, both in the UK and EEC is summarised.*

INTRODUCTION

Vitamin C (L-ascorbic acid) was not isolated in a pure form until 1932 by Szent-Györgyi, but the therapeutic properties of the previously unidentified substance had been recognised since 1545. In 1753 Captain James Lind, physician to the British Fleet, published his paper on the successful treatment of scurvy by the supplementation of the diet with oranges and lemons. This led ultimately to the issue of lime juice, another citrus juice, to men of the British navy.

Thus we can appreciate that vitamin C has had many years of association with fruit juices, and this has led not unnaturally to the inclusion of vitamin C in certain soft drink formulations.

In addition to the use of vitamin C for the nutritional improvement of such products, ascorbic acid also has important applications in the stabilisation of the flavour of fruit juices and soft drinks and in the processing of fruit juices. There are, however, certain problems associated with the naturally occurring or added vitamin C in such products.

NATURAL OCCURRENCE OF VITAMIN C

Vitamin C is soluble in water and is quite widely distributed in nature, being found in many fruits especially the citrus, in green leafy vegetables and in potatoes. The richest natural sources of vitamin C are the Acerola or West Indian Cherry, and the Rose-hip, although only the latter has found significant commercial exploitation in the UK. Table 1 shows typical concentrations of vitamin C found in a selection of fruits and vegetables.

TABLE 1

Concentrations of ascorbic acid found in fruit and vegetables

	Ascorbic acid (mg/100 g)
Acerola (West Indian Cherry)	1 000–3 000
Rose-hip	10–4 800
Blackcurrant	90–360
Parsley	154
Cabbage—raw	70
—cooked	16
Orange	30–65
Lime	39–62
Lemon	30–55
Grapefruit	37–50
Watercress	60
Tomato	10–40
Apple	1–35
Potato—cooked	2–16

The concentration of vitamin C found in the various fruits can vary considerably, being a function of such diverse factors as country of origin, agronomic practices, variety, age of fruit after harvesting, etc. Thus it may be unreasonable to expect any particular fruit to contain a well defined quantity of the vitamin, especially after it has been subjected to further processing to extract the juice and then prepare a formulated product.

NUTRITIONAL FORTIFICATION WITH VITAMIN C

Vitamin C is essential to man, who is unable to synthesise or store any significant quantity within the body and it is necessary that the diet contains a regular, adequate supply of this vitamin. In 1953, the Medical Research Council[1] recommended an adult daily allowance of 30 mg vitamin C. A more recent publication[2] recommends dietary levels for the various ages and status of individuals as shown in Table 2.

It is interesting to note that the UK recommendations are lower than for some other countries of the world. They ensure a sufficiently high intake of vitamin C to prevent signs of deficiency with an adequate safety margin, rather than aiming for tissue levels over

TABLE 2

Recommended daily intake of vitamin C (D.H.S.S. Report No. 120, 1969, H.M.S.O.)

Age/Status	Ascorbic acid (mg)
Boys and girls	
0 up to 1 year	15
1 up to 9 years	20
9 up to 15 years	25
15 up to 18 years	30
Men	
18 and over	30
Women	
18 and over	30
Pregnancy and lacattion	60

and above saturation point. These differences are illustrated in Table 3.

In the UK about half the adult recommended daily intake (RDI) has traditionally been supplied by fresh potatoes, the remainder coming from fruit and green vegetables. In these days of greater weight consciousness, however, potatoes are one of the first items to be cut out of the diet. Wide seasonal variation in the vitamin C content of potatoes and the greater use of convenience processed potatoes (canned and dried) suggests that they are no longer a reliable source of vitamin C for many people. Vitamin C is added to some instant potatoes.

TABLE 3

Recommended daily intake of vitamin C in various countries

Country	Ascorbic acid (mg)
UK	30
Australia	30
Canada	30
Norway	30
USA	70
W. Germany	75
Switzerland	75
USSR	100

Blackcurrant and rose-hip syrups were developed in the UK in the 1930s as they represented rich, indigenous sources of vitamin C. Demand far exceeded the availability of these fruits and it was necessary to supplement supplies with imported orange juice concentrate. Such products are a convenient way of administering vitamin C to small children whose diets may be insufficiently varied to provide for the MDR. Typical products contain between 60 and 100 mg/fl oz vitamin C and thus the MDR for a child is contained in a relatively small daily dose. Products containing blackcurrant juice have the advantage of enhanced stability of the vitamin C due to the presence of naturally occurring flavonoids derived from the pigments in the fruit. Those products containing rose-hip extract usually have a fairly bland flavour which may make them more acceptable to infants who have less developed tastes. Such products usually contain between 13 and 25% overage (intentional excess) of vitamin C to ensure that the claimed level is maintained throughout the normal shelf-life which may be as long as twelve months. In order to achieve a high level of vitamin C it is generally necessary to fortify the product with synthetic vitamin C to accommodate both the natural variation that is experienced with fruits, and also the effect of some form of processing.

The presence of a high level of vitamin C can have undesirable side effects. The stability of some artificial colours which may be used in formulated products can be impaired due to the strongly reducing conditions, which may often result in bleaching of the added colour. Vitamin C is also involved in non-enzymic browning reactions which result in the formation of coloured melanoidins.[3] The combined effect of these two reactions is to enhance the colour changes which take place in the end product.

When very high levels of up to 1 g/fl oz of vitamin C are considered in a liquid medium, other adverse effects are often observed. Provided that the available water exceeds 4%, decomposition of vitamin C takes place and the carbon dioxide (CO_2) produced results in an increase of pressure within the container. There is also the probability of crystals of oxalic acid or its less soluble calcium salt forming. The decomposition of vitamin C is shown in Fig. 1.

It is to be noted that the first step of this reaction is reversible forming a Redox system, and that dehydro-ascorbic acid has approximately 80% of the antiscorbutic properties of the parent substance. Little dehydro-ascorbic acid is found in fruit juices, however, as the further decomposition to diketogulonic acid proceeds quite rapidly.

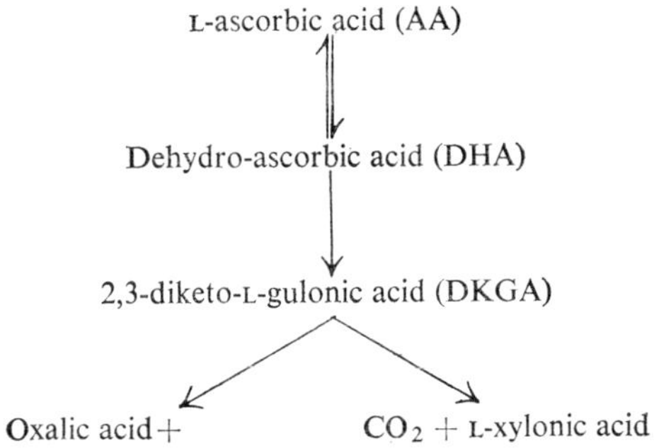

FIG. 1. The decomposition of vitamin C.

VITAMIN C AS A PROCESSING AID

Ascorbic acid finds an important application in the extraction of juice from apples. The first step of a typical juice extraction process is to coarsely mill the washed fruit, and it is at this stage that marked browning may occur due to the oxidation of the naturally occurring tannins by the action of the enzyme polyphenol oxidase. The addition of a small quantity of vitamin C (*ca.* 150 mg/kg) will inhibit the action of the polyphenol oxidase on the tannins, as the ascorbic acid will be preferentially oxidised.[4] The peroxidase enzymes will also degrade ascorbic acid and may be successfully inhibited by the addition of a low level (70 mg/kg) of sulphur dioxide, thus enabling the ascorbic acid to retain its maximum effect. It is thus possible to minimise the undesirable changes to both colour and flavour of juice which would occur if the natural fruit enzymes were allowed to proceed unhindered. This procedure may also be adopted in bottled apple juice, especially when there is an intermediate processing of juice to concentrate which would reduce the inhibiting effect of sulphur dioxide (SO_2).

VITAMIN C AS AN ANTIOXIDANT

In small bottles of fruit juice with a high headspace-to-contents ratio, the addition of a low level of ascorbic acid will significantly improve the colour and flavour stability of the product on the shelf. About 15 mg/fl oz is generally quite adequate to achieve this effect.

Certain types of soft drink are particularly prone to light induced oxidation of the flavour constituents, this applying especially to orange and grapefruit products, based on either fruit or essences. These changes are due to the presence in the product of unsaturated terpenes which are highly reactive, and which produce the characteristic rancid, oily off-notes when exposed to even indirect light of significant intensity. The addition of between 5 and 15 mg/fl oz of ascorbic acid, depending on whether the product is ready-to-drink or requiring dilution respectively, can exert a marked stabilising effect on the flavour. The mode of action is not clearly understood but it is likely that the ascorbic acid utilises any dissolved oxygen. Such addition can, however, have undesirable side effects especially in the presence of added artificial colours when accelerated 'fading' of the colour may be experienced. It is therefore necessary to strike the optimum balance between improved flavour stability and reduced colour stability to suit the particular needs of each product. This can best be established by conducting a series of storage tests under the appropriate conditions when selected combinations of ascorbic acid and artificial colour can be compared. This effect can also be achieved by the use of iso-ascorbic acid, which whilst having only one-twentieth of the antiscorbutic potency, has the same antioxidant properties. The salts of ascorbic acid may be also used in an anti-oxidant role.

THE STABILITY OF VITAMIN C

Various papers reporting the stability of vitamin C in fruit juices and soft drinks have been published in the past.[5–7] It is generally recognised that the main cause of vitamin C loss is due to oxidation under aerobic conditions, but this does not preclude anaerobic decomposition which may also result in many undesirable side effects.

In fruit juices whether packed as single strength natural juices, or reconstituted from a juice concentrate, losses of vitamin C begin even before actual processing begins. Significant losses can occur during any post-harvest storage period particularly in over-ripe and damaged fruit, where enzyme induced oxidative changes are of importance. For the maintenance of a maximum level of vitamin C it is essential to ensure that the fruit is picked at optimal maturity when the vitamin C content is at a maximum and the fruit is in the best physical condition to withstand storage. Once the fruit has entered the process chain, it is important to complete the operations in the shortest possible time, thus reducing to a minimum exposure of the fruit or juice to atmospheric oxygen especially at elevated temperature. The heat inactivation of oxidative enzyme systems at any early stage is very important. Choice of plant construction materials and the maintenance of such equipment in the best state of repair are also important, as aerobic oxidation is significantly increased by trace contamination with metal ions, especially copper and to a lesser extent iron.[8] The use of vacuum deaeration can play a valuable role in vitamin C retention by reducing oxygen content, and a side benefit is a reduction in frothing during any later concentration process which will help to reduce the total process time.

If vacuum deaeration is not possible, then the use of a blanket of carbon dioxide or nitrogen will help to displace air.

The decomposition of vitamin C is accelerated by the presence of fructose and fructose phosphates.[9] It can be expected that as fruit juices are concentrated, or when such concentrates or derived products are subjected to storage, losses of vitamin C will be greater due to the additional fructose produced by inversion of sucrose. In strongly acid solutions (pH 2·2–3·0), vitamin C decomposes to furfural and carbon dioxide.

As has already been mentioned, apple juice contains enzyme systems which are (*inter alia*) responsible for oxidation of vitamin C. The majority of English apples are relatively low in ascorbic acid oxidase, but do contain peroxidases which can be inhibited by the addition of a low level of sulphur dioxide. The ascorbic acid retention will thus be improved adding to the nutritional value of the juice. The presence of ascorbic acid will also inhibit the polyphenol oxidases and reduce changes in colour and flavour. In those apple juices that are naturally low in ascorbic acid, the use of both added SO_2 and ascorbic acid can be particularly valuable.

Blackcurrant juice on the other hand, is rich in flavonoids which exert an antioxidant effect on the vitamin C present. Quercetin has been reported to be particularly effective in this respect in the presence of copper ions.[10] As blackcurrant juice represents a rich indigenous source of vitamin C in the UK considerable work has been carried out to determine the optimum storage conditions to preserve its vitamin C content. Being a seasonal product it is necessary to store large quantities of juice as concentrate for production of blackcurrant syrup throughout the year. It has been found that a 5 or 6 times concentrate can be stored for up to 4 years at 4°C under an atmosphere of carbon dioxide, with only minimal losses of vitamin C.

Formulated soft drinks present quite a different situation as regards vitamin C stability. They may contain fruit juice as the source of some or all of the vitamin C, but in addition contain other substances, *e.g.* sugar, acids, flavours, etc., which may have an adverse effect on vitamin C stability. When a nutritional claim is being made for a specified minimum quantity of vitamin C in a product, the effect of these other substances is of particular importance as they will determine the overage required to support the label claim throughout the expected shelf-life of the product. The use of sulphur dioxide as a preservative in such products is of considerable value as this will have the effect of improving vitamin C stability. A side effect of SO_2 is the deterioration in the stability of any natural pigments or added colouring. The other preservative currently permitted in soft drinks in the UK, benzoic acid, has much less effect on colour stability, but has no stabilising effect on the vitamin C content. It is thus general practice to either use SO_2 alone for maximum vitamin C stability and accept the colour changes that occur in the product, or to use a mixed preservative system where colour stability is of prime importance or the flavour is not compatible with a high level of SO_2. In such cases it will be necessary to allow even greater overage on the vitamin C content.

Several novel techniques for improving the stability of vitamin C have been published in recent years. The addition of a mixture of glucose-oxidase and catalase can be effective in removing oxygen from the system provided that a trace of glucose is present as a substrate. This system is also reported to improve flavour and colour stability.[11] The addition of glucose-oxidase on its own has the effect of increasing vitamin C losses.[12] The incorporation of carefully controlled quantities of amino acids such as cysteine,[13] glycine, histidine and methionine[14] is reported as having a similar stabilising effect on vitamin C, without affecting flavour at the recommended levels of use.

The stability of vitamin C is also determined to a considerable extent by the nature of the packaging used. By virtue of their design, glass bottles have traditionally had a considerable headspace above the contents. This cannot be easily avoided in those products which are pasteurised in bottle, as room has to be allowed for the natural expansion of the product. The headspace will usually contain atmospheric oxygen especially in the case of non-carbonated drinks, which will facilitate aerobic oxidation of the vitamin C. Carbonated products are better in this respect, as the presence of CO_2 will reduce the oxygen content of the product by the effect of its high partial pressure, and careful design of filling equipment can ensure the maximum displacement of oxygen from the headspace. Despite some reports to the contrary, losses of vitamin C in products can also be expected to be greater on exposure to light, especially in the presence of oxygen. This effect may not be due to the direct exposure of the vitamin C to light, but more probably is the photo-induced reaction of vitamin C with other constituents of the product. This feature becomes of greater importance in ullaged bottles where there is additional oxygen present as a result of the enlarged headspace.[15]

It is invariably found that losses of vitamin C increase with increasing temperature and this must be taken into account when formulating vitamin C fortified soft drinks and fruit juice products, to ensure that adequate overage is provided. With the rapid growth of supermarkets where the temperature can average between 60 and 65°F this can present an additional problem to the manufacturer who has to ensure that the product always meets any label claim for vitamin C.

In soft drinks packed in the conventional tinplate can, albeit internally lacquered, losses of vitamin C can occur. There are conflicting reasons for this, one suggesting that the oxygen in the headspace is directly responsible, the other suggesting that the oxygen is first involved in corrosion of the tinplate and that the resulting ferric ions then catalyse vitamin C decomposition. There is general agreement, however, that after a sharp initial loss of vitamin C subsequent stability is quite good. In our experience vitamin C has been shown to be involved in complex reactions with constituents of fruit ingredients and added colouring, to accelerate can corrosion and produce undesirable off-flavours. It may well be that the advent of the aluminium drawn and wall-ironed can will result in improved vitamin C stability presenting greater opportunities for the vitamin C fortification of canned fruit juices and soft drinks.

The majority of plastic materials that have been used for the packaging of soft drinks to date, for example polyethylene and PVC, do not have particularly good barrier properties to oxygen permeability. The shelf-life of soft drinks of this type of pack is relatively short in comparison to the same product packed in glass, being reduced by about 50%. The main deterioration that occurs is the development of off-flavours due to the oxidation of flavour constituents. Vitamin C has not found much application in such products because of the rapid losses that occur by oxidation. More recently, improved plastics using new polymers, co-extrusion of selected laminates, or the use of coating materials having greater barrier properties, *e.g.* PVDC, have resulted in containers giving a shelf-life comparable to glass. Some plastics are now even capable of containing carbonated beverages and vitamin C stability can be expected to be much improved when using such materials.

In powder form soft drink preparations vitamin C stability is generally very good because of the low available water content. Such products can even be packed in sachets provided that a laminate is chosen which reduces moisture permeability to a minimum.

ANALYTICAL QUALITY CONTROL OF VITAMIN C

In those products where vitamin C is added to improve nutritional value, to act as an antioxidant, or as a processing aid, there is a need to determine vitamin C concentration. When a label claim for vitamin C is made, the manufacturer has a legal obligation to ensure that this claim is met throughout the shelf-life that can be reasonably expected of the product under normal trade and in-home conditions. When ascorbic acid is added as an antioxidant or processing aid it is important to ensure that the correct quantity is added, as too much or too little can have undesirable effects. There is also the economic factor to be considered.

Ascorbic acid has the property of reducing the dye phenolindo-2,6-dichlorophenol and the classical technique for estimating vitamin C utilises this feature. Fruit juices and soft drinks which are not very highly coloured can be analysed by direct titration of an aliquot part containing an approximately known quantity of ascorbic acid, with a standardised solution of the indophenol dye. The end point of the titration is a faint pink coloration which persists for 15 sec.[16]

In very highly coloured samples a visual end-point is not practical and a more sophisticated technique must be used. We have adopted a modified electrometric technique using a twin platinum electrode,[17] which is very suitable for routine quality control purposes and gives quick and reliable results.

The presence of sulphur dioxide interferes with the reaction between the indophenol dye and ascorbic acid and it is necessary to first treat the sample with acetone to yield a final concentration of about 20%, and to keep the treated sample in the dark for 3–4 min before titration.

The determination of ascorbic acid may be further complicated by the presence of a group of substances called 'reductones'. These react with the indophenol dye in the same way as ascorbic acid but have no antiscorbutic properties. The 'reductones' in fruit juices are not true reductones, but are probably formed as a result of heat treatment or storage of the juice in the presence of pectin substances, which yield reductic acid when heated in an acid environment.[19]

The 'reductones' may be estimated by first condensing the ascorbic acid with formaldehyde at pH 2 and then carrying out the normal indophenol titration when the 'apparent' vitamin C due to 'reductones' can be determined.[20] This figure can then be subtracted from the 'total apparent' vitamin C to give the actual ascorbic acid content.

Various other substances may also interfere with the determination of ascorbic acid, *e.g.* sulphydryl compounds and traces of reduced iron and tin. Techniques for overcoming this interference have been developed.[21,22]

LEGISLATION

In the United Kingdom, vitamin C is a permitted food additive as prescribed by The Soft Drink Regulations, 1964.[23] There are no limits defined, but a Code of Practice drawn up in 1949 specified the levels necessary to support any nutritional label claim for the presence of vitamin C. The normal serving of product had to contain at least one sixth of the adult MDR for the presence of vitamin C to be claimed at all, or one half of MDR for the product to be called a 'rich source' of vitamin C. This Code of Practice has now been largely superseded by the Labelling of Food Regulations, 1970,[24] which permit claims to be made provided that the *actual* quantity of vitamin C present is declared. Both L-ascorbic acid and dehydroascorbic acid are considered to be equivalent in this respect, and they may be referred to as either vitamin C or ascorbic acid, although the quantity must be specified in terms of milligrammes of ascorbic acid. Within the EEC, regulations currently vary from country to country, but the draft Directives for both Fruit Juices and Non-alcoholic Refreshing Beverages allow the use of ascorbic acid as a processing aid only. There is no provision made for therapeutic claims based on the presence of vitamin C.

CONCLUSION

This paper has attempted to review the information which is available concerning the application of vitamin C in soft drink and fruit juice manufacture. With an ever increasing number of suggested therapeutic applications for this vitamin, it is possible that fortified soft drinks and fruit juices could play an important role in dietary supplementation with vitamin C.

ACKNOWLEDGEMENTS

The author would like to thank colleagues in the Research and Development Department for their assistance in preparing this paper, and The Beecham Group Ltd for permission to publish it.

REFERENCES

1. Bartley, W., Krebs, H. A. and O'Brien, J. R. P. (1953). *Spec. Rep. Ser. Med. Res. Coun., Lond.*, **280,** H.M.S.O., London.
2. *Recommended Intakes of Nutrients for the United Kingdom* (1969), Rep. on Pub. Health and Med. Subj., **120,** H.M.S.O., London.
3. Clegg, K. M. (1966). *J. Sci. Fd Agric.*, **17**(12), p. 546.
4. Kuusi, T. and Pajunen, E. (1971). *Maataloustieteellinen Aikakauskirja,* **43**(1), p. 20.
5. Bender, A. E. (1958). *J. Sci. Fd Agric.*, **9,** p. 754.
6. Pelletier, O. and Morrison, A. B. (1965). *J. Amer. Diet. Assoc.*, **47**(5), p. 401.
7. Wells, J. O. (1966). *Can. Fd Ind.*, **37**(11), p. 43.
8. Timberlake, C. F. (1960). *J. Sci. Fd Agric.*, **11**(5), p. 268.
9. Huelin, F. E. (1953). *Food Res.*, **18,** p. 633.
10. Harper, K. A., Morton, A. D. and Rolfe, E. J. (1969). *J. Fd Technol.*, **4,** p. 255.
11. Scott, D. (1967). *Am. Soft Drink J.*, **121**(859), p. 27.
12. Tschogowadse, S. K. and Bakuradse, N. S. (1972). *Lebensmittel-Industrie*, **19**(7), p. 287.
13. Merck & Co. Inc. (1972). *British Patent* 1,277,393.
14. Merck & Co. Inc. (1972). *U.S. Patent* 3,652,290.
15. Hellström, V. and Andersson, R. (1969). *Var Föda* (1), p. 5.
16. *Official Methods of Analysis of the Association of Official Analytical Chemists* (1970). 11th Edn. p. 777. *A.O.A.C.* Washington, D.C. 20044.
17. Liebmann, H. and Ayres, A. D. (1954). *Analyst*, **70,** p. 411.
18. Mapson, L. W. (1942). *Biochem. J.*, **36,** p. 196.
19. Wokes, F. and Organ, J. G. (1944). *Qu. J. Pharm. Pharmacol.*, **17**(3), p. 188.
20. Kuusi, T. (1960). *Suomen Kemistiechti B*, **33,** p. 139.
21. Mapson, L. W. (1943). *J. Soc. Chem. Ind.*, **62,** p. 223.
22. Pelletier, O. and Morrison, A. B. (1966). *J.A.O.A.C.*, **49**(4), p. 800.
23. *Statutory Instrument* 1964 *No.* 760 *as amended* 1969 *No.* 1818, H.M.S.O., London.
24. *Statutory Instrument* 1970 *No.* 400 *as amended* 1972 *No.* 1510, H.M.S.O., London.

DISCUSSION

Goodman: At one point you said that SO_2 increases colour loss, and at another you said that it prevents it. Do you have any figures correlating colour change with nutritional value?
Gresswell: Sulphur dioxide can stabilise the vitamin C, but will reduce colour stability. There is a subtle balance between SO_2 and vitamin C for optimal flavour and colour stability.
Degkwitz: You gave the breakdown of ascorbic acid to dehydro-ascorbic acid as one step. As far as I know, it is more complicated than that. Only one electron is removed so semi-dehydro-ascorbic acid remains. Two of these radicals disproportionate then to 1 AA and 1 DHA molecule.
Gresswell: I will take what you say to be the absolute truth. I presented the popular conception of the decomposition of vitamin C.
Degkwitz: There is a difficulty about these recommendations for *daily* intake of AA, since it is known that blood levels are normalised already in the evening after a single intake in the morning. So if you split the amount recommended into one part for the morning and one for the evening, it will be of more benefit than taking the whole amount at once.
Gresswell: I think this is true. All the evidence suggests that spreading vitamin C over daily food intake is better than a single dose, as the body can only absorb at a certain rate and any excess is excreted.
Cooke: How does optimum ripeness with regard to vitamin C content relate to optimum ripeness as far as extraction (in terms of yield, colour, flavour, etc.) is concerned?
Gresswell: In blackcurrants, vitamin C content is maximum when the fruit is at optimum ripeness for processing. Before optimum maturity, vitamin C and fruit juice yield are lower. Beyond optimum maturity, damage to the fruit structure results in enzymic and oxidative loss of vitamin C, and the fruit is more difficult to process.
Long: Will vitamin C prevent the loss of colour with azo dyes?
Gresswell: The stability of some azo dyes in canned soft drinks in the presence of vitamin C will depend very much on the formulation. The presence of certain fruit constituents can increase losses of colour by complex interaction with vitamin C.
Goodman: Do you have any information correlating colour change with nutritional value?
Gresswell: No.
Wilson: Have you got any information as to whether, if there is a colour change or degeneration, it has any toxic effect?
Gresswell: The toxicity of azo dyes is well covered, but little is known about the toxicity of breakdown products.
Chairman: Certain approved food colours, known as 'Heinz body-formers', on reductive cleavage of the azo-link, form metabolites which produce haematological effects. If ascorbic acid brings about this reductive cleavage, then it can be postulated that in certain cases the derivations of colour bleaching would produce measurable haematological effects.

11

Vitamin C in Breadmaking

B. H. Thewlis

Flour Milling & Baking Research Association, Chorleywood, Rickmansworth, Herts, England

ABSTRACT

Vitamin C is widely used as an additive to dough in the course of making bread. During mixing, it is believed that oxidation takes place to give dehydro-L-ascorbic acid, and that this then oxidises the sulphydryl groups of flour proteins, thus changing the physical properties of the dough in such a way that, after baking, a good loaf is produced. The ultimate fate of the vitamin C during this process is of interest, and it has been shown, using radioactive techniques, that about 24% of the carbon present in it is lost, mostly and probably entirely, as carbon dioxide. The rest of the carbon remains in the bread as a mixture of water-soluble acidic substances whose major component appears to be L-threonic acid; this accounts for about 52% of the carbon in the added ascorbic acid. Lesser amounts of other products present include 2,3-diketo-L-gulonic acid. None of the original L-ascorbic acid, or of its likely decomposition products, dehydro-L-ascorbic acid and oxalic acid, can be detected in the bread.

INTRODUCTION

History of the Use of Vitamin C in Breadmaking

It has long been known that the texture, eating and keeping properties of bread are made better, and the loaf volume is increased, if small amounts of certain substances, known as improvers, are added to the flour. It was at first thought that only oxidising agents such as potassium bromate possessed this property; however, in 1935 Jørgensen[1] reported that vitamin C (L-ascorbic acid) gave a similar improving action. Since then, L-ascorbic acid has become widely used as an additive to flour used for breadmaking, both in the EEC, where in most cases it is the only improver that is permitted, and in the UK, where it is commonly used in conjunction with potassium bromate. More than 70% of British bread is now made from dough containing L-ascorbic acid, one of its main advantages being its tolerance to overtreatment. An addition of 75 ppm based on flour used has been found to be adequate for all flours likely to be encountered.

Reactions Taking Place in Dough

Jørgensen thought that the improving action of L-ascorbic acid was due to its conversion to dehydro-L-ascorbic acid, which then inhibited proteases present in the flour. As a result of diminished protein hydrolysis, there was greater gas-retaining capacity of the dough proteins. This theory has now been modified in order to take account of the sulphydryl–disulphide relationships existing in dough; cysteine occurs about once in every 1200–1600 amino acids of protein chains and cystine at about forty times this frequency, so that the following well known type of interchange reaction can occur between neighbouring protein chains:

$$RSSR + R'SH \rightarrow RSSR' + RSH$$

When the dough ingredients (flour, water, salt, yeast) are first brought together, the protein molecules absorb water and begin to swell and uncoil, so that reactive groups on different chains approach each other, and the likelihood of their interaction increases. In this way, protein molecules which were originally distinct, become linked into the vast network known as gluten. On standing, the process continues with the effect that disulphide bonds are broken and reestablished in different positions, so that the overall effect is a gradual expansion and spreading out of the network. In the older process of 'bulk fermentation', this could take about 3 hr; however, in modern 'mechanical development' processes it is greatly accelerated by very intense mixing, which is believed to break some disulphide bonds by a mechanochemical reaction, thus speeding the rearrangement of the protein chains. In either process a point is reached when the dough is 'ripe', *i.e.* it has reached its optimum condition for gas-holding, and the reaction must then be stopped. It is here that dehydro-L-ascorbic acid from the added vitamin C is believed to play its part, by increasingly oxidising protein sulphydryl groups and thereby taking them out of circulation, so that the interchange reactions cease. At the same time, the final protein network is stabilised by the formation of disulphide cross-links.

The details of the reaction have been investigated by a number of authors. Meredith[2] reported that wheat flour contained ascorbic oxidase, although oxidation of added L-ascorbic acid was only partly due to this, and there was no evidence that the rest was due to metal iron catalysis. Honold and Stahmann[3] could find very little ascorbic oxidase activity in their flour; the possibility remains, therefore, that oxidation may take place through some other enzyme system. However, it is well established that wheat flour contains an active dehydro-ascorbic acid reductase,[4] which can mediate the oxidation of thiol groups (*see* Fig. 1). Neither the oxidative nor reductive stages are limiting for dough improvement.[3]

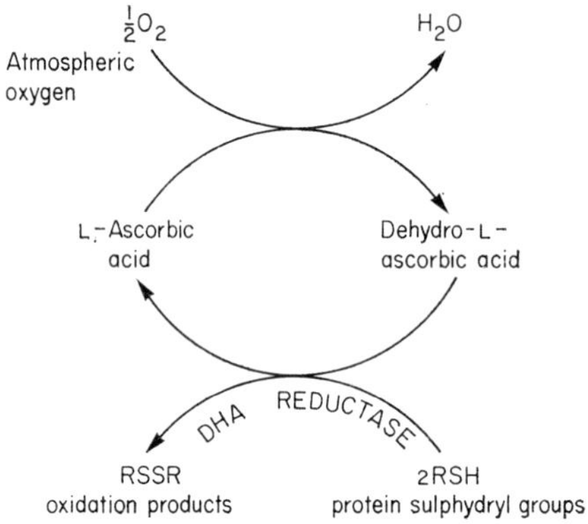

FIG. 1.

These reactions have been shown to take place in dough and flour–water suspensions at ordinary temperatures; however, further changes might be expected to take place in the later fermentation and baking stages. Since no information appeared to be available on this subject, it was decided to investigate the ultimate fate of L-ascorbic acid in a baked loaf, with particular reference to the possible formation of harmful substances.

THE ULTIMATE FATE OF L-ASCORBIC ACID IN BREADMAKING

L-ascorbic acid is known to be highly reactive, and capable of decomposing in a number of different ways. Thus in aqueous solution at room temperature the reaction shown as ‘Route 1’ is known to occur[5] (*see* Fig. 2).

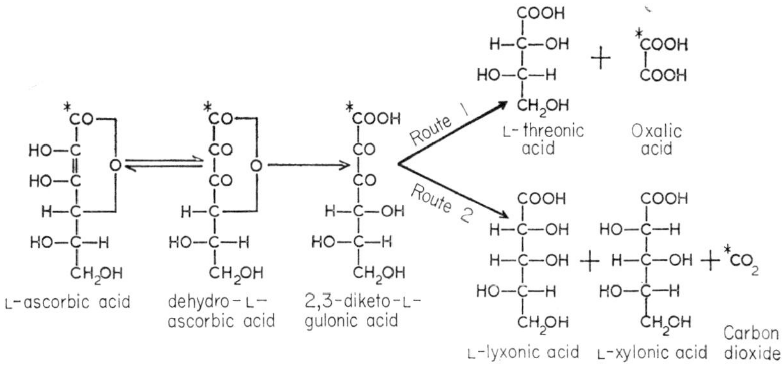

FIG. 2.

It was shown in a recent study[6] that carbon dioxide was also evolved during this process, indicating that a second type of decomposition was taking place simultaneously, and 'Route 2' was postulated. Other workers[7] have shown that decarboxylation can proceed along the entire carbon atom chain of L-ascorbic acid under certain circumstances. These reactions take place at ordinary temperatures, but still further changes are possible on heating; thus, fifteen different products were isolated[7] from a boiled solution of L-ascorbic acid in water, of which the major one was furfural, previously detected by other workers. Any or all of these substances might, therefore, be present in a loaf made from flour containing L-ascorbic acid. Since bread is a complex mixture and the amount of L-ascorbic acid added is small it is evident that decomposition products can only be identified if isotopically labelled L-ascorbic acid is used.

Baking

The baking equipment was designed for collecting and estimating all volatile radioactive products. Mixing, moulding and fermentation took place in a glove box, and baking in a special air-tight oven. In each case issuing gases were passed through a cold trap followed by an absorber containing a mixture of 2-phenylethylamine (30 ml) and 2-ethoxyethanol (30 ml), to retain any $^{14}CO_2$.

In the baking procedure adopted, flour (140 g) and fat (1 g) were mixed with water (80 ml) containing ascorbic acid (10·5 mg), salt (2·5 g) and yeast (3 g). After mixing and moulding, the dough was rested at 43°C for 10 min, re-moulded and then allowed to ferment at 43°C for 45 min. The loaf was baked at 220°C for 24 min; during this time air was drawn through as rapidly as possible to carry volatile products over into the absorption system. When baking was complete the loaf had to be taken out of the oven, and any remaining volatile products were lost. In spite of various modifications of technique, the added radioactivity could not be completely accounted for, probably owing to losses of $^{14}CO_2$. This was also the experience of other workers.[8] On cooling, the crust and crumb were separated, dried and powdered. Samples were combusted in oxygen, and the specific radioactivity of each fraction was determined in a liquid scintillation counter. Further details of experimental methods have been published elsewhere.[9]

Use of Terminally Labelled L-ascorbic Acid

Terminally labelled L-[1-^{14}C] ascorbic acid (for labelling *see* asterisks in Fig. 2) was obtainable commercially. When it was added to dough, there was practically no loss of radioactivity during mixing and fermentation of the dough; however, during baking the $^{14}CO_2$ evolved contained from 27 to 31% of the radioactivity originally added, there being no other volatile product. This figure for carbon dioxide was probably lower than the true value, owing to the unavoidable losses already referred to, but showed that some decomposition of L-ascorbic acid took place in baking. The residual radioactivity of the bread was associated almost entirely with its aqueous extract, which was found to contain one main unknown radioactive substance (Table 1).

There was, therefore, no evidence for the presence of any L-ascorbic

TABLE 1

R_f values of unknown substance derived from L-[1-^{14}C] *ascorbic acid, and of reference substances*

Unknown	System A 0·11	System B 0·05
L-ascorbic acid	0·42	0·25
Dehydro-L-ascorbic acid	0·52	0·37
2,3-diketo-L-gulonic acid	0·11	0·06
Oxalic acid	0·64	0·70

System A = Paper chromatography using 1-butanol–acetic acid–water (4:1:5 by vol).
System B = Thin layer chromatography using acetonitrile–butyronitrile–water (66:33:2 by vol.)

acid or dehydro-L-ascorbic acid; the main product appeared to be 2,3-diketo-L-gulonic acid. Further conversion of this might have been expected, along one of the paths in Fig. 2; thus, Route 1 would give oxalic acid, and it was particularly desirable to ascertain whether or not this had been produced, in view of its toxic nature. However, the above chromatograms gave no indication of the presence of oxalic acid; extracts tested further for oxalate by addition of calcium chloride in the presence of inactive oxalic acid to act as a carrier gave a precipitate containing no radioactivity, and the total radioactivity of the filtrate remained unchanged. Dough to which [U-^{14}C] oxalic acid had been added evolved only 0·4% of the radioactivity as $^{14}CO_2$ during baking, showing that there had been no transient formation and decarboxylation of oxalic acid under the conditions of baking. L-Ascorbic acid had evidently undergone decarboxylation at the C1 position, with formation of a non-volatile and non-radioactive product which remained in the loaf. Route 2 suggested that this might be a mixture of L-lyxonic and L-xylonic acids; to obtain further information, it was necessary to use uniformly labelled L-ascorbic acid.

Use of Uniformly Labelled L-ascorbic Acid

L-[U-^{14}C] ascorbic acid was prepared by the method of Frush and Isbell[10] from L-[U-^{14}C] sorbose, which was obtainable from the Radiochemical Centre, Amersham. When used in baking, it gave the results shown in Table 2.

TABLE 2

Radioactivity distribution during breadmaking with addition of L-[U-^{14}C] *ascorbic acid to the dough*

Fraction	% added radioactivity found Bake 1	Bake 2	Bake 3
Fermentation condensate	0	0	0
Fermentation carbon dioxide	0·5	1·9	1·8
Baking condensate	0	0	0
Baking carbon dioxide	5·6	5·2	4·2
Bread (crust + crumb)	79·4	73·5	76·0
Recovery total	85·5	80·6	82·0
Specific activity of crumb	0·58	0·54	0·58
Specific activity of crust	0·60	0·56	0·57

TABLE 3

Aqueous extraction of bread made using L-[U-^{14}C] *ascorbic acid*

	Bake 1	Bake 2	Bake 3
% of bread radioactivity extracted by water	99	97	98·5
% of bread radioactivity retained by CG-400 resin column	95	95	93

The only volatile radioactive product that could be detected was $^{14}CO_2$, and this was mostly evolved during baking; as before, the amounts quoted were probably lower than the true values. Other possible volatile products, such as furfural, which would have led to radioactive condensate in the cold traps, were absent; it therefore seemed reasonable to suppose that a more likely value for the

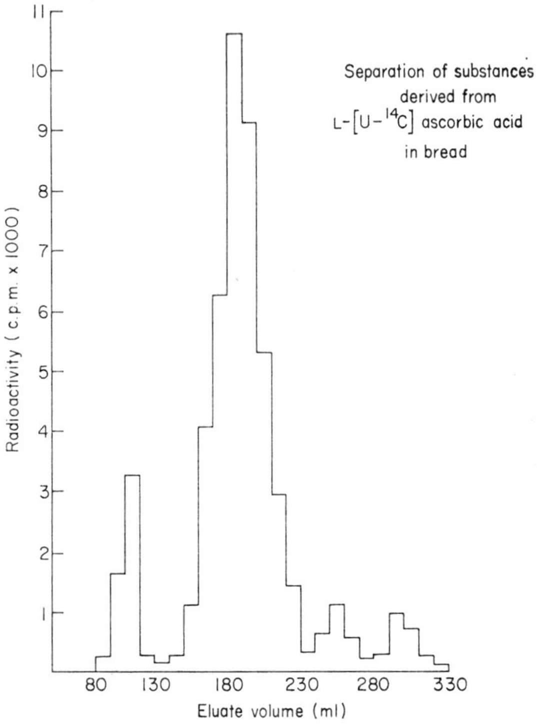

FIG. 3.

$^{14}CO_2$ evolved might be obtained from the residual bread radioactivity by difference, *i.e.* about 24%. Most of the products derived from ascorbic acid remained in the loaf, and specific activities of crumb and crust were about equal, suggesting that decarboxylation had proceeded to about the same extent in each. It was shown that practically all the radioactive material present in the bread was

TABLE 4

R_f values of unknown substance derived from L-[U-^{14}C] *ascorbic acid, and of reference substances, obtained during paper chromatography with the solvent systems shown*

	System				
	A	B	C	D	E
Unknown	0·38	0·19	0·26	0·38	0·12
L-ascorbic acid	0·42	0·27	0·35	0·40	0·40
L-threonic acid	0·38	0·20	0·25	0·38	0·11
L-lyxonic acid	0·37	0·28	0·24	0·43	0·35
L-xylonic acid	0·25	0·23	0·17	0·29	0·22

System A = 1-butanol–acetic acid–water (4:1:5 by vol.).
System B = ethyl acetate–acetic acid–water (3:1:3 by vol.).
System C = water saturated butanol–formic acid (95:5 by vol.).
System D = l-propanol–formic acid–water (6:3:1 by vol.).
System E = 1-butanol–ethanol–water (5:1:4 by vol.).

water-soluble, and also that most of it was acidic in nature (Table 3).

Hence neutral substances, such as furfural or xylosone, could be present in only small amounts as decarboxylation products. The radioactivity could be eluted completely from the column with 0·01 N formic acid, as shown in Fig. 3.

The major component was separated and subjected to paper chromatography using five different systems (*see* Table 4).

In each system, therefore, the unknown major component showed the same R_f as L-threonic acid and differed from the other possible decarboxylation products, L-lyxonic and L-xylonic acids, of Route 2. Identity of the unknown was confirmed by carrier dilution analysis and by co-chromatography, using authentic L-threonic acid. It accounted for about 52% of the radioactivity originally added to the flour. The preceding component was likewise shown to have the chromatographic properties of 2,3-diketo-L-gulonic acid, and to account for about 11% of the initial radioactivity (Table 5).

TABLE 5

Recovery of radioactivity from bread baked using L-[U-^{14}C] *ascorbic acid in the dough*

Radioactivity due to carbon dioxide	24% of original
Radioactivity due to L-threonic acid	52% of original
Radioactivity due to 2,3-diketo-L-gulonic acid	11% of original
Recovery total	87% of original

CONCLUSIONS

It appeared, therefore, that the main breakdown path that was followed by L-ascorbic acid during baking corresponded neither to Route 1 nor to Route 2, but was yet another route involving 2,3-diketo-L-gulonic acid as an unstable intermediate, leading to carbon dioxide and L-threonic acid as the main products. The carbon dioxide did not originate from decarboxylation of oxalic acid.

The literature does not appear to contain any reference to toxicological studies on 2,3-diketo-L-gulonic acid or L-threonic acid as such. However, considerable work has been done on L-ascorbic acid itself in circumstances where these products might be expected to be formed, involving ingestion by humans and by rats, with generally favourable results.[11] The Toxicity Sub-Committee set up to advise on the safety of food additives regards L-ascorbic acid, or vitamin C, as safe for use in breadmaking.

REFERENCES

1. Jørgensen, H. (1935). *Mühlenlaboratorium*, **5,** p. 114.
2. Meredith, P. (1965). *J. Sci. Fd Agric.*, **16,** p. 474.
3. Honold, G. R. and Stahmann, M. A. (1968). *Cereal Chem.*, **45,** p. 99.
4. Carter, J. E. and Pace, J. (1965). *Cereal Chem.*, **42,** p. 201.
5. Herbert, R. W., Hirst, E. L., Percival, E. G. V., Reynolds, R. J. W. and Smith, F. (1933). *J. Chem. Soc.*, p. 1270.
6. Levandowski, N. G., Baker, E. M. and Canham, J. E. (1965). *Biochemistry N.Y.*, **3,** p. 1465.
7. Tatum, J. H., Shaw, P. E. and Berry, R. E. (1969). *J. Agric. Fd Chem.*, **17,** p. 38.

8. Lee, C. C. and Chen, C. H. (1966). *Cereal Chem.*, **43,** p. 695.
9. Thewlis, B. H. (1971). *J. Sci. Fd Agric.*, **22,** p. 15.
10. Frush, H. L. and Isbell, H. S. (1957). *J. Res. Natn. Bur. Stand.* **59,** p. 289.
11. *Tech. Rep. Ser. Wld Hlth Org.* (1962). No. 288, p. 19.

DISCUSSION

Sharman: With regard to the loaf prepared from flour with 75 mg ascorbic acid, you said it would not have been any larger with more ascorbic acid. Would it have done better with a mixture of potassium bromate and ascorbic acid?

Thewlis: Yes . . . most baking flours in fact contain bromate at the beginning and the ascorbate is put in the fat after that. We think the synergism between the two—the bromate plus the ascorbic—is better than the improvement you'd expect from either of them added together.

Sharman: You said that 75% of bread was made in this way. The other 25%—is that mainly chlorine dioxide-treated?

Thewlis: I think it's mainly chlorine dioxide-treated, yes. All 70% extraction flour is. The other is simply made by long fermentation by small bakers who haven't yet got round to this mechano-chemical treatment.

Spencer: Since potassium bromate is not permitted in most European countries, would the level of 150 ppm of ascorbic acid continuously used in those countries produce approximately the same result as the 75 ppm ascorbic acid together with the bromate which is used in the Chorleywood process?

Thewlis: As far as we can judge, yes, but of course their bread is rather different from ours.

Goodman: Why do you have to have the loaf bigger?

Thewlis: Because the housewife likes it that way. If you go to the supermarket and see large loaves and small loaves at the same price, you choose the bigger one, don't you? Also, it is easier to eat—it is less dense.

Goodman: I think you said you used 70% extraction flour. Have you done any of these tests with wholemeal flour?

Thewlis: No, we haven't, because this after all is what most people eat.

Lewin: The loaf on which you use ascorbic acid appeared to me to be lighter in colour. What is the bleaching action due to?

Thewlis: We don't normally obtain any colour effect from ascorbic acid; it may have been simply a lighting effect on the slide.

Spencer: Perhaps I might suggest that the increased volume produced an extended and finer crumb structure which in turn resulted in an apparent whitening effect, due to increased reflection of light?

Lewin: I thought that chlorine dioxide was forbidden now owing to the fact that even dogs go nervous when exposed to it. Does it have no ageing effect?

Thewlis: It is mostly a bleacher. It has some ageing effect.

Lewin: No wonder we are all getting older.
Chairman: Are there any advantages in any other baked products such as biscuits in the use of ascorbic acid?
Thewlis: No; that was a different problem. We put metabisulphite into biscuits.
Hunt: Presumably bread is priced on a weight basis?
Thewlis: Yes. In fact the slides I showed you were somewhat out of date—nowadays the whole thing is very highly automated. It is all done on conveyor bands, to get a one-pound loaf, or slightly above it—as little above as they can, naturally.

12

Role of Vitamin C on Microsomal Cytochromes

E. DEGKWITZ and HJ. STAUDINGER

Biochemical Institute, University of Giessen, Giessen, Germany

ABSTRACT

Investigations of the reaction mechanism of L*-ascorbic acid demonstrated a decrease of the specific activity of the mono-oxygenase to about* 50% *of the starting value in liver microsomes of guinea-pigs deprived of vitamin C for* 14 *days. The activity was measured by the hydroxylation of acetanilide and by the demethylation of aminophenazone and of hexobarbital. The mono-oxygenase activity declines to the same extent as the specific concentration of cytochrome P-*450*, thus showing that* L*-ascorbic acid does not function as an electron donor for the enzyme reaction but influences the concentration of cytochrome P-*450*, the terminal oxidase of the system. The concentration of cytochrome P-*450 *decreases to about* 40% *of the starting value during prolonged vitamin deficiency. The decrease starts within* 24 *hr after omission of vitamin C, probably due to both the rapid decline of* L*-ascorbic acid levels in the liver of guinea-pigs, and to the short half-life time of the cytochrome P-*450*. When liver levels of* L*-ascorbate in guinea-pigs deprived of vitamin C are restored to normal values the concentration of cytochrome P-*450 *returns to normal values within* 24–36 *hr. Administrations of* D*-arabino-ascorbate,* 5*-oxo-*D*-gluconate or δ-aminolevulinic acid or of known inducers of the biosynthesis of the mono-oxygenase like phenobarbital and* 3*-methylcholanthrene, are also able to raise the cytochrome content within this period of time. These results demonstrate that* L*-ascorbic acid is not an essential substrate for the biosynthesis of cytochrome P-*450*. They are consistent with an influence of* L*-ascorbic acid on the biosynthesis of haem. This conclusion is confirmed by the finding that cytochrome b_5, too, decreases in liver microsomes of guinea-pigs deprived of vitamin C. The influence of* L*-ascorbic acid is not limited to the liver since a decrease of cytochromes was also found in spleen, kidney and adrenal microsomes of guinea-pigs during vitamin deficiency. As to standardisation of guinea-pigs, especial care has to be taken of the vitamin C levels in the liver.*

Our research on the influence of L-ascorbic acid on the mono-oxygenase in liver microsomes of guinea-pigs was started in order to obtain further information on molecular reaction mechanisms of vitamin C. Results reported in the literature[1] indicated that L-ascorbic acid influences this enzyme system since less hydroxylated compounds were found, compared to normal, in the urine of guinea-pigs deprived of L-ascorbic acid and treated with acetanilide. The reaction mechanism causing the decrease of the drug metabolism was not yet known.

The mono-oxygenase is connected with cytochrome P-450 as terminal oxidase. It reacts with various drugs, which usually are classified as Type I or as Type II substrates, according to the type of spectral changes of the cytochrome P-450 induced when the substrates are bound to the system.[2] We found a decrease in the mixed function oxygenation of both types of substrates in isolated liver microsomes of guinea-pigs deficient in L-ascorbic acid. The mono-oxygenase activity was tested by N-demethylation of aminophenazone and mixed function oxygenation of hexobarbital, both Type I substrates, and by *p*-hydroxylation of acetanilide, a Type II substrate. The specific activities were decreased to 50–60 % of the starting value in liver microsomes of guinea-pigs deprived of L-ascorbic acid for about 14 days (Table 1). This decrease was paralleled by a higher sensitivity of the guinea-pigs to treatment with drugs. We had to reduce the amount of phenobarbital administered in our experiments, referred to later, to about 60 % of the amount given to normal guinea-pigs, since the deficient animals did not survive otherwise.

In the beginning of L-ascorbic acid deficiency, for about 14 days the decrease of the mixed function oxygenation of all substrates tested was caused by a decline of the V_{max} of the specific overall reaction activity, the apparent Michaelis constant for the substrates remained unaltered. But later on, during long-lasting L-ascorbic acid deficiency we found alterations of kinetic constants for the *p*-hydroxylation of acetanilide. There was an increase in V_{max} and in the apparent Michaelis constant for acetanilide (Table 1).

Since guinea-pigs suffer a decrease in weight if they are deprived of L-ascorbic acid for more than about 14 days, we carried out investigations in animals which had been limited in food intake. The results served as a control to check whether alterations found in guinea-pigs during long-lasting L-ascorbic acid deficiency were due to a lack of the vitamin or only due to their malnutrition. We did not arrange pair feeding for this purpose since we felt that this method implies unaltered absorption of food as well as its utilisation in guinea-pigs deprived of vitamin C and that this has not yet been proved. We limited the amount of food to such a degree that the

TABLE 1

Influence of vitamin C deficiency and of limitation of food uptake on the activity of the mono-oxygenase in liver microsomes of guinea-pigs and on the content of its terminal oxidase cytochrome P-450. All values in bold are significantly different from those of the control animals. Some of these data have already been published in Leber et al.[3] *and by Degkwitz* et al.[4,5]

Supply to guinea-pigs	Cytochrome P-450 specif. conc. (nMol/mg prot.)	Demethylation of aminophenazone[a] specif. activity (mU/mg prot.)	p-Hydroxylation of acetanilide specif. activity (mU/mg prot.) V_{max}	p-Hydroxylation of acetanilide K_M acetanilide (mMol/litre)	Number of animals
Normal diet *ad libitum*[b]	1·43 ± 0·11	21·5 ± 2·2	10·4 ± 0·6	2·0 ± 0·5	10
Diet without L-ascorbic acid:					
for 14 days[c]	**0·82** ± 0·13	**11·5** ± 0·5	**6·2** ± 0·4	2·5 ± 0·7	10
for 21 days[b]	**0·68** ± 0·06	**8·9** ± 0·6	9·3 ± 0·5	**10·0** ± 1·0	10
Normal diet amounts limited for 12 days[c]	1·51 ± 0·08	**35·0** ± 2·3	12·4 ± 0·7	1·5 ± 0·4	8
Normal diet *ad libitum* and treatment with 3-methylcholanthrene[b]	1·56 ± 0·11	18·6 ± 1·7	11·2 ± 0·7	**0·5** ± 0·1	5

[a] The Michaelis constant for this substrate remained unaltered.
[b] *See* Ref. 3 for methods.
[c] *See* Refs. 4 and 5 for methods.

guinea-pigs suffered the same decrease in weight as the animals deprived of L-ascorbic acid. The food amount was distributed in several supplies per day.

The data in Table 1 demonstrate that the alterations of both the apparent kinetic constants for the hydroxylation of acetanilide are not due to the malnutrition of the guinea-pigs. The increase in the V_{max} for the specific activity of N-demethylation of aminophenazone found in guinea-pigs limited in food uptake is paralleled in certain phases of long-lasting L-ascorbic acid deficiency, but in both cases there is no change in the apparent Michaelis constant for this substrate.

Investigations of other enzyme activities in liver microsomes of guinea-pigs deprived of vitamin C gave as a result no alterations specific for the lack of L-ascorbic acid in glucose-6-phosphatase (EC 3.1.3.9), carboxylesterase (EC 3.1.1.1) acting on procaine,[6] and activities of NADPH-oxidation.[6] Thus there is no general decrease of microsomal enzyme activities.

Looking at the components of the mono-oxygenase we found a decrease in the concentration of cytochrome P-450, the terminal oxidase of the system. The decrease conformed with the decrease in the specific enzyme activities for all the substrates tested unless the apparent Michaelis constant for acetanilide was altered. Thus there was no difference in the reaction activity of the enzyme system based on the number of moles of cytochrome present in the microsomes (Table 2).

These results indicated that the decrease in mixed function activity

TABLE 2

Activity of the mono-oxygenase in liver microsomes of normal and deficient guinea-pigs acting on racemic hexobarbital. The values in bold are significantly different from those of the control animals. These data were already published by Gundermann et al.[7]

Supply to guinea-pigs	Mixed function oxygenation of hexobarbital		Number of animals
	specif. activity (mU/mg prot.)	'molecular' activity (mU/nMol cyt. P-450)	
Normal diet	1·06 ± 0·06	0·53 ± 0·03	6
Diet without L-ascorbic acid for 21 days	**0·53** ± 0·05	0·56 ± 0·05	6

found in liver microsomes of guinea-pigs deficient in L-ascorbic acid is due to a decrease of the cytochrome P-450 concentration and not to a lack of the vitamin being involved in the enzyme reaction.

The increase of both apparent kinetic constants for the *p*-hydroxylation of acetanilide in guinea-pigs deprived of the vitamin for a long period of time might have indicated that the vitamin too is involved in the enzyme reaction but that its concentration necessary for this part is so low that an impairment of the enzyme activity does not take place earlier. But a direct involvement of the vitamin in the

TABLE 3

Influence of pretreatment of guinea-pigs deprived of vitamin C for more than 21 *days on the kinetics of p-hydroxylation of acetanilide by the monooxygenase in liver microsomes. All values in bold are significantly different from those of the control animals. For methods see Refs.* 4 *and* 13. 5-*Oxo*-D-*gluconate was administered in the same amounts as* L-*ascorbic acid*

Administration to deficient guinea-pigs	*p*-Hydroxylation of acetanilide K_M acetanilide (mMol/litre)	specif. activity (mU/mg prot) V_{max}	Number of animals
None	10·0 ± 1·0	9·3 ± 0·5	10
L-Ascorbic acid/48 hr	**1·8** ± 0·5	8·9 ± 0·8	6
3-Methylcholanthrene/48 hr	**0·8** ± 0·1	11·4 ± 0·6	5
Dehydro-ascorbic acid/1 hr	**3·3** ± 1·0	**4·2** ± 0·7	6
5-Oxo-D-gluconate/10 min	**5·0** ± 0·5	**4·4** ± 0·4	5
Ethionine/1 hr	**1·5** ± 0·6	**4·6** ± 0·8	4

enzyme reaction can be excluded since it is possible to normalise the kinetics of the action on acetanilide in the liver microsomes of such guinea-pigs. Table 3 presents results of intraperitoneal administrations a short time before killing. Injections of either dehydro-ascorbic acid or of 5-oxo-D-gluconate or even of ethionine equally reduce the apparent Michaelis constant for acetanilide to normal values and cause a decrease of the V_{max} for the specific *p*-hydroxylation activity to about 50%. So that there is no difference any more in the overall reaction activity related to the number of moles of cytochrome P-450 present in the microsomes, as found in the beginning of L-ascorbic acid deficiency. Since Schulze and Staudinger[8] found a change in the relation of lipid and protein contents in the liver microsomes of our guinea-pigs deprived of the vitamin for a long time it seems possible that the corresponding alterations of the membrane structure of the endoplasmic reticulum cause alterations of the kinetics of the mono-oxygenase action on acetanilide.

The enzyme system seems to be more modifiable in its action on this Type II substrate for there is not only a special variability of the apparent Michaelis constant for acetanilide in guinea-pigs deprived of L-ascorbic acid but also in normal animals. Treating normal guinea-pigs with 3-methyl-cholanthrene causes a decrease of the apparent Michaelis constant for acetanilide only, not for the other substrates (Table 1).

After arriving at the result that the decrease in the activity of the drug metabolising mono-oxygenase in livers of guinea-pigs deprived of L-ascorbic acid is due to a decrease of the amount of the terminal oxidase cytochrome P-450 present in the microsomes and not due to a lack of the vitamin involved in the enzyme reaction, we tried to obtain information on the reaction mechanism causing the decrease of the cytochrome concentrations. But first we had to look into the course of the vitamin C levels in the livers of guinea-pigs supplied with L-ascorbic acid by different methods. In estimating the L-ascorbic acid levels of our guinea-pigs limited in food uptake we had observed that the vitamin C levels of guinea-pig livers are able to change rather quickly.

The data in Table 4 demonstrate a decrease of the L-ascorbic acid levels in the guinea-pig livers to about 60 % of the starting value after 14 hr of fasting only (*i.e.* by withdrawal of food overnight). There were no changes of the vitamin levels in the spleen and in the adrenals.

These results explained our former difficulties of keeping up normal vitamin levels in the livers of guinea-pigs limited in food uptake. Pharmacokinetics of L-ascorbic acid seem to differ in the

TABLE 4

Influence of fasting of guinea-pigs on the L-ascorbic acid in liver, spleen and adrenals. All values in bold are significantly different from those of the control animals

Supply to guinea-pigs	L-ascorbic acid levels (μg/g organ, fresh weight) liver	spleen	adrenals	Number of animals
Diet *ad libitum*	300 ± 75[a]	480 ± 40	1 730 ± 160[a]	20
Diet deprived for 14 hr	**175** ± 40[a]	430 ± 40	1 400 ± 140[a]	8

[a] These data were already published by Degkwitz *et al.*[5]

liver, from other organs. We had no further difficulties since we administered L-ascorbic acid several times per day instead of once a day.

Figure 1 demonstrates that the L-ascorbic acid levels in the liver of guinea-pigs supplied with the vitamin are only raised for a few hours by a single additional supply via stomach tube. This result is

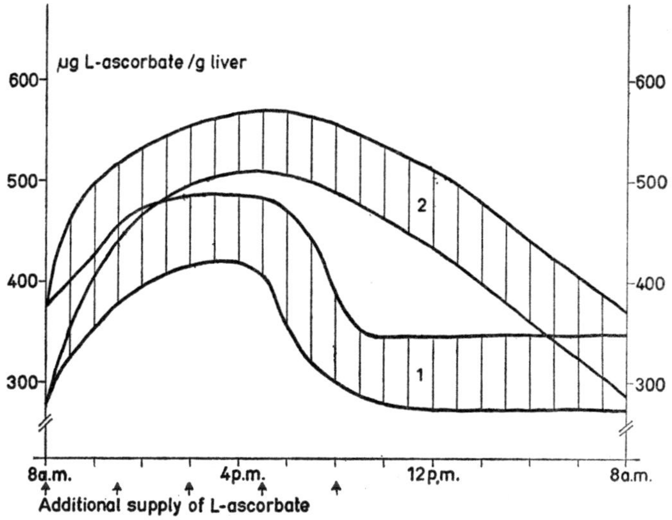

FIG. 1. Ranges of L-ascorbic acid levels in the livers of guinea-pigs fed *ad libitum* a diet containing 0·68% (w/w) vitamin C and additionally provided with L-ascorbate by stomach tube once (area 1) or several times, at the times indicated (area 2). Area 2 is based on data partly already published by Degkwitz and Kim.[9]

in agreement with that of Table 4. The further data show that it is hardly possible to raise the L-ascorbic acid levels in guinea-pig liver above concentrations of 300–400 µg vitamin C per g of fresh weight, even by increasing the supplies via mouth.

The results presented in Fig. 2 indicate that the inefficiency of increased supply *per os* is not due to tissue saturation in the liver, for they are easily raised to concentrations of 800–900 µg/g, fresh weight, by intraperitoneal injections. This outcome must be due to a limited absorption capacity for vitamin C in the guinea-pig. The same is already known for man.[10] Based upon these results it seems more reasonable that the so-called phenomenon of tissue saturation

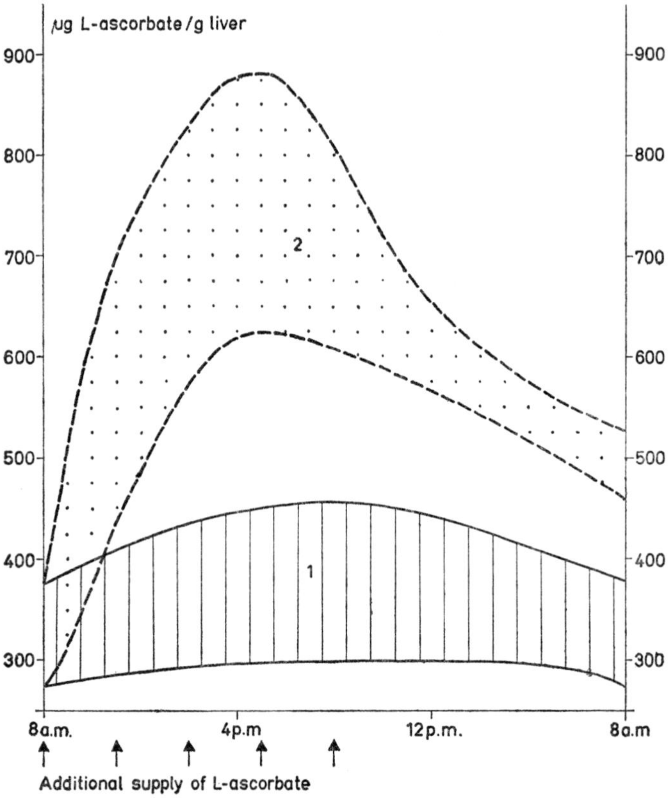

FIG. 2. Ranges of L-ascorbic acid levels in the liver of guinea-pigs fed *ad libitum* a diet containing 0·68% (w/w) vitamin C and additionally provided with (intraperitoneal) injections of dehydro-ascorbic acid once (area 1) or with L-ascorbate several times, at the times indicated (area 2). Area 2 is based on data partly already published by Degkwitz and Kim.[9] Dehydro-ascorbic acid was administered in the same amounts as L-ascorbate.

is due to substrate saturation of the enzyme systems responsible for the absorption of the vitamin.

The second curve in Fig. 2 presents the results of an additional intraperitoneal injection of dehydro-L-ascorbic acid to guinea-pigs. Dehydro-L-ascorbic acid is often thought to be the general transport form of the vitamin in the organism. But, it does not raise the vitamin levels of the liver to equal concentrations like L-ascorbic acid.

Figure 3 demonstrates the corresponding L-ascorbic acid levels in the spleen and in the adrenals. They seem far more stable in the spleen than in the liver and there is no significant dependence on the administration method. The vitamin C levels of the adrenals are not significantly raised by additional supply via stomach tube, but they were decreased for several hours by intraperitoneal injections. The decline is intensified by administration of D-arabino-ascorbic acid instead of L-ascorbic acid.

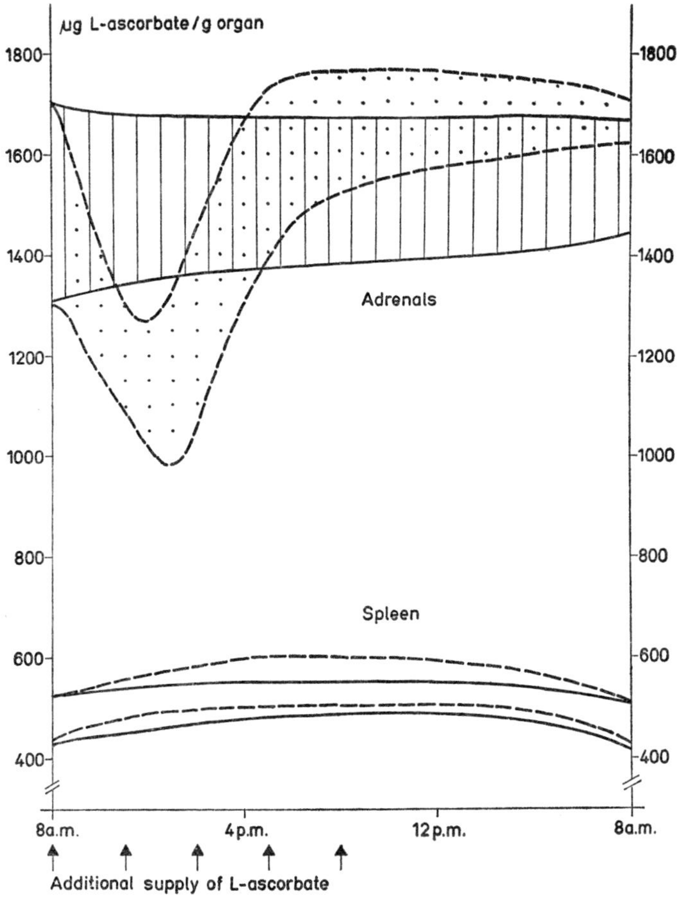

FIG. 3. Ranges of L-ascorbic acid levels in the spleen and in the adrenals of guinea-pigs fed *ad libitum* a diet containing 0·68% (w/w) vitamin C and additionally provided with L-ascorbic acid by intraperitoneal injections (area dotted) or by stomach tube (area hatched) at the times indicated.

The results presented in Figs. 1–3 imply several consequences. Their common unawareness of each others' techniques probably explains many differences in the results published by other authors upon the influence of vitamin C. It is insufficient simply to state the amount of vitamin C administered per day since at least some of the resulting L-ascorbic acid levels depend on the administration manner. The finding of normal contents of L-ascorbic acid in some organs may not be valid as well for the liver concentrations. Estimation of the L-ascorbic acid levels in the liver only allow conclusions on the concentrations during a few previous hours. Besides, resaturation of liver levels to concentrations of 300–400 μg/g of fresh weight in deficient guinea-pigs succeeds about 10 hr faster by intraperitoneal injections.[9]

According to these results we supplied our guinea-pigs *ad libitum* with a diet containing 0·68 % (w/w) of vitamin C for at least 14 days in order to standardise their L-ascorbic acid levels. Since guinea-pigs eat haphazardly and their absorption capacity for vitamin C is limited we obtained low variation of the L-ascorbic acid levels of our animals around the clock, even in the liver.

This concentration of vitamin C might be surprising, but the data presented in Table 5 show that L-ascorbic acid levels of the animals tend to deviate to a greater extent as soon as the vitamin content of the diet is reduced. The result might be explained by differences in the amount of food uptake already influencing the amount of L-ascorbic acid absorbed.

The period of time necessary for standardisation of guinea-pigs depends on the investigations in view. It has to be considered that supplying a diet completed with high contents of vitamin C and

TABLE 5

Influence of vitamin C concentration in the diet fed ad libitum *to guinea-pigs on the L-ascorbic acid levels in liver and adrenals. For methods see Ref.* 5

Vitamin C content of the diet	Ranges of L-ascorbic acid levels (μg/g organ, fresh weight) liver	adrenals	Number of animals
0·60 % (w/w)	275–395	1230–1940	5
0·50 % (w/w)	165–350	1070–1570	5
0·33 % (w/w)	115–320	1350–1990	5
0·09 % (w/w)	90–300	500–2100	5

with a mixture of minerals easily absorbed by the intestine has to be extended to more than 4 weeks in order to re-obtain normal concentrations of non-haem iron in the liver, the spleen and in the blood serum of the guinea-pigs, for these concentrations rise significantly in the meantime as demonstrated for the blood serum levels in Table 6.

TABLE 6

Increase of non-haem iron levels in blood serum of guinea-pigs supplied ad libitum *with a diet completed with a mixture of minerals and containing* 0·68% (w/w) *of vitamin C. All values in bold are significantly different from those of the animals supplied with standard diet. The iron levels of blood serum were measured according to Deggau* et al.[11] *For further methods see Ref.* 5

Supply to guinea-pigs	Non-haem iron levels in blood serum (μg/100 ml)	Number of animals
Standard diet	277 ± 35	9
Special diet		
for 7 days	**341** ± 16	5
for 14 days	**431** ± 19	5
for 21 days	**339** ± 54	10
for 28 days	212 ± 14	5

Our investigations upon correlations between the L-ascorbic acid levels and the concentrations of cytochrome P-450 in the liver gave as a result no increase of the amounts of cytochrome by raising the vitamin levels above concentrations of 300–400 μg/g fresh weight. A decline of the L-ascorbic acid levels in the liver to about 50% of the starting value caused by omission of the vitamin supply to the guinea-pigs for 24 hr led to a significant decrease of the cytochrome P-450 concentrations as presented in Fig. 4. The rapid decline of the cytochrome P-450 concentration is of course due to the lucky fact that this haemoprotein has a very short half-life.

If withdrawal of the vitamin is continued the L-ascorbic acid levels of the liver decline to values less than 2%, but the concentrations of cytochrome P-450 do not decrease to less than about 40% of the starting value, even during long-lasting L-ascorbic acid deficiency (Fig. 4).

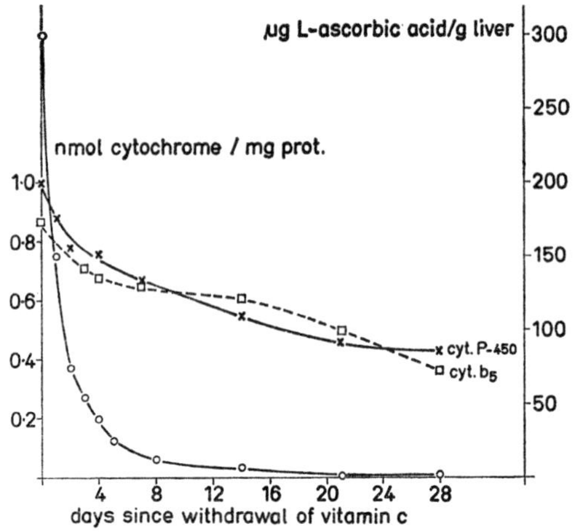

FIG. 4. Course of L-ascorbic acid levels and of specific concentrations of microsomal cytochromes P-450 and b_5 in the livers of guinea-pigs after withdrawl of the vitamin. These data were already published by Degkwitz *et al.*[5]

Table 7 presents the results of our investigations on possible ways to restore the cytochrome P-450 concentration to normal in deficient guinea-pigs. The amounts of cytochrome P-450 could be raised within 48 hr by intraperitoneal injections of L-ascorbic acid. In subsequent studies we found normal values within 24–36 hr after elevation of the vitamin levels to 300–400 μg/g of liver, fresh

TABLE 7

Influence of protein biosynthesis inhibitors on the effect of L-*ascorbic acid on the concentrations of cytochrome P*-450 *in liver microsomes of guinea-pigs deficient in the vitamin for* 14 *days. All values in bold are significantly different from those of the controls. Most of these data were already published by L. Höchli-Kaufmann.*[12] *All substances were administered by intraperitoneal injections*

Administration to deficient guinea-pigs	Cytochrome P-450 specif. conc. (nMol/mg prot.)	Number of animals
None	0·63 ± 0·07	10
L-ascorbic acid	**0·88** ± 0·07	6
L-ascorbic acid and actinomycin D	0·67 ± 0·09	4
L-ascorbic acid and ethionine	0·63 ± 0·06	6
L-ascorbic acid and cycloheximide	0·70 ± 0·13	4

weight.[9] Hence decline and normalisation of the cytochrome P-450 concentration succeed in about the same period of time. The effect of L-ascorbic acid could be prevented by simultaneous injections of inhibitors of protein biosynthesis, like actinomycin D, cycloheximide or ethionine.

Table 8 demonstrates that normalisation of the cytochrome P-450 concentration could also be produced by treatment of the deficient guinea-pigs with drugs known to induce the biosynthesis of the mono-oxygenase, phenobarbital and 3-methylcholanthrene. It seems interesting that additional injections of vitamin C increased the induction of the cytochrome by phenobarbital. This indicates that the reaction mechanism of vitamin C must be different from that of the barbital. Phenobarbital is known[14–20] to stimulate both haem biosyntheses by induction of the biosynthesis of δ-amino-levulinic acid synthetase and of ferrochelatase (EC 4.99.1.1), and biosynthesis of the apoprotein, and also induces an increase of the amount of the endoplasmic reticulum in the liver cell.[15]

Since normal metabolism of a haemprotein like cytochrome P-450 presupposes intact biosynthesis of the haem and of the apoprotein, the results presented in Tables 7 and 8 did not yet permit

TABLE 8

Influence of treatment with phenobarbital and with 3-*methylcholanthrene for* 48 *hr on the concentration of cytochrome P*-450 *in liver microsomes of normal guinea-pigs and of animals deficient in the vitamin for more than* 21 *days. All values in bold are significantly different from those of the corresponding controls. These data were already published by Leber* et al.[13] *All substances were administered by intraperitoneal injections*

Administration to deficient guinea-pigs	Cytochrome P-450 specif. conc. (nMol/mg prot.)	Number of animals
(a) Normal guinea-pigs		
None	1·43 ± 0·11	15
3-Methylcholanthrene	1·56 ± 0·10	6
Phenobarbital	**2·32** ± 0·20	6
(b) Deficient guinea-pigs		
None	0·68 ± 0·06	10
3-Methylcholanthrene	**1·26** ± 0·05	5
Phenobarbital	**1·36** ± 0·05	5
Phenobarbital and L-ascorbic acid	**2·42** ± 0·11	5

a differentiation between an action of the vitamin on haem or on protein biosynthesis. Accordingly we looked for an adequate experiment, enabling us to distinguish between these two possibilities. We tested the influence of δ-aminolevulinic acid, the product of the rate limiting step of haem biosynthesis, and we found an increase in the concentration of cytochrome P-450 by intraperitoneal injections within 48 hr as presented in Table 9. Thus the decrease of the cytochrome P-450 concentration in the livers of guinea-pigs deficient in vitamin C seems to be due to an influence of L-ascorbic acid upon the metabolism of haem.

TABLE 9

Influence of treatment with δ-aminolevulinic acid on the concentrations of cytochromes P-450 *and* b_5 *in liver microsomes of guinea-pigs deficient in the vitamin for* 21 *days. All values in bold are significantly different from those of the controls. These data were already published by Luft* et al.[21] *All substances were administered by intraperitoneal injections*

Administration to deficient guinea-pigs	Specific cytochrome concentration (nMol/ mg prot.) cyt. P-450	cyt. b_5	Number of animals
None	0·46 ± 0·11	0·47 ± 0·12	12
δ-aminolevulinic acid	**0·90** ± 0·16	**0·65** ± 0·12	6
δ-aminolevulinic acid and L-ascorbic acid	**0·99** ± 0·14	0·54 ± 0·08	6
δ-aminolevulinic acid and actinomycin D	0·60 ± 0·07	0·46 ± 0·03	7

This conclusion is confirmed by the finding that there is not only a decrease of the cytochrome P-450 concentration but also a decrease of the amounts of microsomal cytochrome b_5 in the livers of guinea-pigs deprived of the vitamin (Fig. 4). The concentrations of cytochrome b_5, too, decrease to about 40% of the starting value and are increased in deficient guinea-pigs as well by injections of L-ascorbic acid as of δ-aminolevulinic acid within 48 hr. Both decrease and increase occur more slowly compared to those of the cytochrome P-450. The delay is demonstrated in Table 10. It seems to conform to the longer half-life of cytochrome b_5.[22–24]

Table 10 additionally represents results of our investigations carried out in order to discover if substances bearing structural analogy to L-ascorbic acid have the same influence on the cytochrome

TABLE 10

Influence of 5-oxo-D-gluconate and of D-arabino-ascorbate on the concentrations of cytochromes P-450 and b_5 in liver microsomes of guinea-pigs deprived of the vitamin for 14 days in comparison to the effect of L-ascorbic acid. All values in bold are significantly different from those of the controls. These data were already published by Degkwitz et al.[9] *All substances were administered by stomach tube*

Administration to deficient guinea-pigs	Specific cytochrome concentration (nMol/mg prot.) cyt. P-450	cyt. b_5	Number of animals
None	0·65 ± 0·06	0·58 ± 0·04	10
L-Ascorbate 5×/24 hr	**0·82** ± 0·09	0·57 ± 0·09	6
L-ascorbate 15×/72 hr	**1·52** ± 0·18	**1·07** ± 0·13	5
5-Oxo-D-gluconate 5×/24 hr	**0·89** ± 0·09	**0·77** ± 0·14	5
D-arabino-ascorbate 15×/72 hr	**0·86** ± 0·18	0·55 ± 0·12	9

contents. D-arabino-ascorbic acid yields the same basic results, but research on the influence of this substance is difficult since it is nearly impossible to obtain equally high concentrations of D-arabino-ascorbic acid as by administration of L-ascorbic acid, especially in the liver of guinea-pigs.

The result that 5-oxo-D-gluconate has the same effects as L-ascorbic acid seems very interesting, although the reaction mechanism has not yet been proved to be identical. The structure of 5-oxo-D-gluconate is depicted in Fig. 5. The substance does not function as an oxidation–reduction system like L-ascorbic acid and D-arabino-ascorbic acid, it does not react with dinitrophenylhydrazone and is not metabolised to a substance reacting with dinitrophenylhydrazone in the guinea-pig liver. The results indicate that L-ascorbic acid might

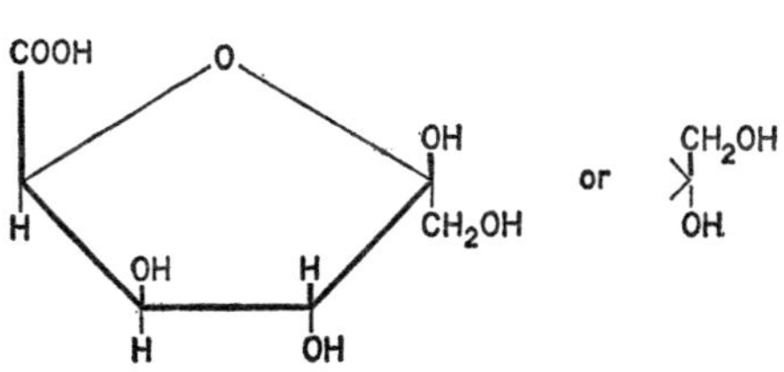

FIG. 5. Structure of 5-oxo-D-gluconate.

function by other than oxidation–reduction reaction influencing the cytochrome contents. Until now vitamin C has been regarded mainly as an essential electron donor for certain hydroxylation reactions, although the standard potential of the system semidehydro-ascorbate/L-ascorbate, E'_0 (pH 7·0) = 0·34 V,[25] being rather high does not favour such a role.

We hope that further investigations on the influence of L-ascorbic acid and of 5-oxo-D-gluconate may help to distinguish the reaction mechanisms of the vitamin. The results obtained so far indicate that 5-oxo-D-gluconate prevents decreases of the cytochrome concentrations and restores normal amounts in deficient guinea-pigs, but it does not seem to substitute for the vitamin's influence on collagen metabolism, since the guinea-pigs continue to suffer haemorrhages.

We also measured the amounts of microsomal cytochromes in other organs of normal and deficient guinea-pigs in order to investigate whether the influence of the vitamin is limited to the liver. The results indicated corresponding decreases in the concentrations of the microsomal cytochromes P-450 and b_5 in the adrenals and in the kidneys and of the microsomal cytochrome b_5 in the spleen. (We did not estimate the cytochrome P-450 concentrations in this organ since there are difficulties with the high amounts of haemoglobin.)

We do not yet know the reaction mechanism of the influence of L-ascorbic acid. We thoroughly investigated non-haem iron levels of normal and deficient guinea-pigs since differences in the concentration of storage iron were reported[26] and it was tempting to conclude an influence via iron metabolism. But we did not find any correlation between the concentrations of non-haem iron and the amounts of microsomal cytochromes in the liver or in the spleen or in the kidneys. There are alterations in the concentrations of storage iron in the liver and in the spleen of guinea-pigs deprived of L-ascorbic acid. But they depend on the period of time passed since the withdrawal of the vitamin and mainly seem due to haemorrhages which start rather early[27] during L-ascorbic acid deficiency and increase until the death of the guinea-pigs.

ACKNOWLEDGEMENT

The support of the Deutsche Forschungsgemeinschaft for these investigations is gratefully acknowledged.

REFERENCES

1. Axelrod, J., Udenfried, S. and Brodie, B. B. (1954). *J. Pharmacol. exp. Therapeut.*, **111,** p. 176.
2. Remmer, H., Schenkman, J., Estabrook, R. W., Sasame, H., Gillette, J., Narasimhulu, S., Cooper, D. J. and Rosenthal, O. (1966). *Mol. Pharmacol.*, **2,** p. 187.
3. Leber, H.-W., Degkwitz, E. and Staudinger, Hj. (1969). *Hoppe-Seyler's Z. Physiol. Chem.*, **350,** p. 439.
4. Degkwitz, E., Leber, H.-W., Kaufmann, L. and Staudinger, Hj. (1970). *Hoppe-Seyler's Z. Physiol. Chem.*, **351,** p. 397.
5. Degkwitz, E., Höchli-Kaufmann, L., Luft, D. and Staudinger, Hj. (1972). *Hoppe-Seyler's Z. Physiol. Chem.*, **353,** p. 1023.
6. Degkwitz, E., Luft, D., Pfeiffer, U. and Staudinger, Hj. (1968). *Hoppe-Seyler's Z. Physiol. Chem.*, **349,** p. 465.
7. Gundermann, K., Degkwitz, E. and Staudinger, Hj. (1973). *Hoppe-Seyler's Z. Physiol. Chem.*, **354,** p. 238.
8. Schulze, H.-U. and Staudinger, Hj. (1970). *Hoppe-Seyler's Z. Physiol. Chem.*, p. 184.
9. Degkwitz, E. and Kim, K. S. (1973). *Hoppe-Seyler's Z. Physiol. Chem.*, **354,** p. 555.
10. Kübler, W. and Gehler, J. (1970). *Int. Z. Vitaminforsch.*, **40,** p. 442.
11. Deggau, E., Kröhnke, F., Schnalke, K. E., Staudinger, Hj. and Weis, W. (1965). *Zeitschr. f. Klin. Chem.*, **3,** p. 102.
12. Höchli-Kaufmann, L. (1971). *Dissertat. Univ. Gießen.*
13. Leber, H.-W., Degkwitz, E. and Staudinger, Hj. (1970). *Hoppe-Seyler's Z. Physiol. Chem.*, **351,** p. 995.
14. Remmer, H. (1962). In *Ciba Found. Sympos. on Enzymes and Drug Action* (Eds. Mongal, J. L. and Reuck, A. V. S. de). S276, Little Brown, Boston.
15. Remmer, H. and Merker, H.-J. (1963). *Klin. Wochenschr.*, **41,** p. 276.
16. Remmer, H. and Siegert, M. (1964). *Naunyn-Schmiedebergs Arch. Exp. Pharmakol. Pathol.*, **247,** p. 522.
17. Granick, S. (1966). *J. Biol. Chem.*, **241,** p. 1359.
18. Marver, H. S., Schmid, R. and Schützel, H. (1968). *Biochem. Biophys. Res. Commun.*, **33,** p. 969.
19. Baron, J. and Tephly, T. R. (1970). *Arch. Biochem. Biophys.*, **139,** p. 410.
20. Hasegawa, E., Smith, C. and Tephly, T. R. (1970). *Biochem. Biophys. Res. Commun.*, **40,** p. 517.
21. Luft, D., Degkwitz, E., Höchli-Kaufmann, L. and Staudinger, Hj. (1972). *Hoppe-Seyler's Z. Physiol. Chem.*, **353,** p. 1420.
22. Omura, T., Siekevitz, P. and Palade, G. E. (1967). *J. Biol. Chem.*, **242,** p. 2389.
23. Bock, K. W. and Siekevitz, P. (1970). *Biochem. Biophys. Res. Commun.*, **41,** p. 374.
24. Greim, H., Schenkman, J. B., Klotzbücher, M. and Remmer, H. (1970). *Biochim. Biophys. Acta*, **201,** p. 20.

25. Everling, F. B., Weis, W. and Staudinger, Hj. (1969). *Hoppe-Seyler's Z. Physiol. Chem.*, **350,** p. 886.
26. Lipschitz, D. A., Bothwell, P. H., Seftel, H. C., Wapnick, A. A. and Charlton, R. W. (1971). *Brit. J. Haematol.*, **20,** p. 155.
27. Gore, J., Wada, M. and Goodman, M. L. (1968). *Arch. Path.*, **85,** p. 493.

DISCUSSION

Hornig: Is it possible to normalise cytochrome P-450 through vitamin C?

Degkwitz: It doesn't normalise, because we were not able to get normal ascorbic acid levels. If you take deficient animals and feed them ascorbic acid you are able to raise P-450, not to normal levels, but to a significant extent.

13

Vitamin C in Lipid Metabolism and Atherosclerosis

E. GINTER

Institute of Human Nutrition Research,
Bratislava, Czechoslovakia

ABSTRACT

In guinea-pigs and primates with acute scurvy numerous disorders of lipid metabolism are described: changed cholesterol concentrations in blood and tissues, increased serum β-lipoproteins, increased cholesterol synthesis, decreased oxidation of fatty acids and cholesterol, decreased mobilisation of fatty acids, etc. Acute scurvy is, metabolically, a very complicated state and it is probable that most of these disorders are of secondary character. Moreover the model of acute scurvy is unrealistic in regard to vitamin C deficiency in a human population.

A new model of chronic latent vitamin C deficiency, 'hypovitaminosis C', in guinea-pigs is developed. The model is based on the administration of a scorbutogenic diet for 14 *days and consequent administration of a maintenance dose of ascorbic acid of* 0·5 *mg per animal per* 24 *hr. Chronic hypovitaminosis C significantly increases total cholesterol concentration in the blood serum, liver and skin. In guinea-pigs fed an atherogenic diet with* 0·3 *per cent of cholesterol, hypovitaminosis C causes an increased cholesterol accumulation in many tissues, including thoracic aorta. Research on the mechanism of hypovitaminosis C intervention in cholesterol metabolism performed by isotope technique, absorption of cholesterol-*4-^{14}C, *cholesterol biosynthesis from acetate-*1-^{14}C, *transformation of cholesterol-*4-^{14}C *to* ^{14}C*-bile acids, oxidation of cholesterol-26-*^{14}C *to* $^{14}CO_2$, *and two-pool analysis of plasma cholesterol die-away curves, demonstrates a significantly lowered conversion of cholesterol to bile acids in guinea-pigs with chronic latent vitamin C deficiency. Decreased cholesterol catabolism in hypovitaminous guinea-pigs causes hypercholesterolaemia, prolongs half-life of plasma cholesterol and, in some animals, causes atheromatous changes in thoracic aorta and coronary arteries.*

INTRODUCTION

Considerable discrepancies exist as regard views on the role of vitamin C in lipid metabolism and atherogenesis. An attempt has therefore been made to analyse the prevailing notions and also to present our own concept of the function of ascorbic acid in cholesterol metabolism and atherogenesis.

LIPID METABOLISM IN ACUTE SCURVY

Scores of works have been written describing various disorders of lipid metabolism in guinea-pigs and monkeys, brought about by experimentally induced scurvy. Detailed reviews of these may be found in monographs.[1,2] Scurvy is regularly accompanied by fatty infiltration of the liver and increased β-lipoprotein concentration in blood serum.[3-5] But data on the effect of acute avitaminosis C on the cholesterol levels in plasma and in various tissues are contradictory in the extreme.[2] The great majority of authors, however, agree that cholesterol is accumulated in the body during acute avitaminosis C.[6,7] A probable cause of this phenomenon is an enhanced cholesterol synthesis from acetate[8,9] and a slowed down cholesterol catabolism.[10] Attempts by certain authors to ascribe cholesterol accumulation in the body of scorbutic animals to a lowered cholesterol transformation to steroid hormones do not seem plausible. The function of ascorbic acid in this process has not as yet been elucidated[11,12] and the amount of the cholesterol metabolised in this way is negligible.[13] Of interest are data on a lowered mobilisation of non-esterified fatty acids[14] and on a slowed down oxidation of fatty acids[15] in scorbutic animals, which should be further processed.

An interpretation of biochemical data obtained in scorbutic animals is not easy, for metabolically scurvy presents a very complicated state. Thus, for instance, disorders of cholesterol metabolism are probably only secondary sequelae to an impaired function of the tricarboxylic acids cycle in the liver of scorbutic animals,[16] resulting in increased utilisation of the acetate pool for cholesterol synthesis.[2,9] It appears probable that further disorders of lipid metabolism reported in scorbutic animals are also related to a total metabolic disorganisation provoked by an abrupt drop in body weight, haemorrhages, a negative nitrogen balance, etc.[2]

VITAMIN C AND EXPERIMENTAL ATHEROSCLEROSIS

Considerable discrepancies exist likewise in this field.[2] According to some authors,[17-19] high doses of ascorbic acid depress alimentary hypercholesterolaemia in rabbits and rats and exert a protective effect as regards experimental atherosclerosis. Other authors, on the other hand, reject such an action of vitamin C.[20-23] As our own results with rabbits were not unambiguous,[24] recently we followed the effect of ascorbic acid on cholesterol turnover in rabbits fed an atherogenic diet to which 0·3% of cholesterol had been added. In three groups of male rabbits (controls; atherogenic diet; atherogenic diet + 0·2% ascorbic acid) we followed die-away curves of plasma cholesterol specific activity after a one pulse labelling with 4-^{14}C-cholesterol. The results of mathematical analysis in terms of a two-pool model[25] are summarised in Table 1.

Experimental atherosclerosis provoked a significant enlargement of both cholesterol pools and acceleration of cholesterol turnover rate in the rabbits regardless of whether these had received zero or high doses of ascorbic acid. Removal of cholesterol from the circulation was slowed down and the values of the rate constant for an irreversible excretion of cholesterol and its catabolites from the organism were also lowered. Pool and kinetic parameters found in rabbits with a zero and a high ascorbic acid intake proved to be practically even. Likewise, the course of hypercholesterolaemia and cholesterol accumulation in the liver, adrenals and thoracic aorta showed a similar pattern in the two groups. In keeping with these biochemical findings, the degree of atheromatous changes of the aorta and coronary arteries matched in both the groups.

These data, however, do not permit us to conclude that ascorbic acid fails to affect experimental atherosclerosis. Neither the rabbit, nor the rat is a suitable animal for these kind of studies. These species, in contrast to primates and guinea-pigs, synthesise ascorbic acid and their tissues are saturated with vitamin C independently of the amount of ascorbic acid supplied through diet. In the above study on rabbits we also followed vitamin C concentration in their liver, spleen, adrenals and small intestine, and found practically no difference between the group with a zero and the one with a high intake of vitamin C. It is thus obvious why exogenous vitamin C did not exert any effect on cholesterol turnover and experimental atherosclerosis at all.

More interesting results were obtained on guinea-pigs, *i.e.* an animal species which, like man, is dependent on a supply of exogenous ascorbic acid. Certain authors[5,26] have found early atheromatous lesions and subendothelial lipid deposits in the aorta of guinea-pigs with acute vitamin C deficiency. Others[27-29] failed to detect atheromatous changes in such animals, but have described nuclear

TABLE 1

Size of pools and kinetic parameters of cholesterol turnover in rabbits with alimentary hypercholesterolaemia and atherosclerosis on zero and high intake of ascorbic acid. In the table means from 5 to 7 animals ± S.E.M. are given

	Parameter	Control	Experimental atherosclerosis	
			Vitamin C : 0	Vitamin C : 0·2% in diet
$t_{1/2\alpha}$	Half-life of first exponential (days)	1·6 ± 0·1	2·1 ± 0·2	2·0 ± 0·2
$t_{1/2\beta}$	Half-life of second exponential (days)	19·0 ± 1·2	26·8 ± 3·2	27·8 ± 3·1
M_A	Size of pool A (mg)	1 113 ± 90	4 174 ± 451	3 900 ± 406
$M_{B\,min}$	Minimum size of pool B (mg)	1 635 ± 40	4 392 ± 688	4 569 ± 715
PR_A	Production rate in pool A = turnover rate (mg per 24 hr)	124 ± 7	254 ± 27	247 ± 32
k_A	rate constant for irreversible excretion from pool A	0·116 ± 0·012	0·061 ± 0·001	0·063 ± 0·004

In both groups with experimental atherosclerosis half-life of second exponential is prolonged ($P < 0{\cdot}05$), size of both pools ($P < 0{\cdot}001$) and turnover rate ($P < 0{\cdot}002$) are significantly increased; rate constant for irreversible excretion is significantly decreased ($P < 0{\cdot}002$). There is no difference between the atherosclerotic groups on zero or high intake of vitamin C.

abnormalities of aorta endothelium, separation of endothelial intercellular junction, depletion of collagen, increased hyaluronic acid and decreased chondroitin sulphate B in the aorta of scorbutic guinea-pigs.[27,28] It would appear that the development of acute avitaminosis C in guinea-pigs is too rapid for any atheromatous changes to become evident.

MODEL OF CHRONIC LATENT VITAMIN C DEFICIENCY

For reasons stated above (scurvy as a poorly defined metabolic state, the need of long-term experiments to follow atherogenesis) the need was felt for fresh methodical approaches in dealing with the role of vitamin C in lipid metabolism and atherogenesis. We have therefore designed a model of chronic latent vitamin C deficiency[30] and a model of alimentary hypercholesterolaemia and atherosclerosis[31,32] in guinea-pigs. These models, in contrast to those currently employed, enable the effect of ascorbic acid deficiency on lipid metabolism and atherogenesis to be followed in long-term experiments.

Following a series of preliminary trials, we have adopted the following procedure: guinea-pigs are fed a scorbutogenic diet[33] without addition of any ascorbic acid for a fortnight, during which their body pool of vitamin C declines abruptly, although the vitamin C deficiency does not become clearly manifest. After this time, the animals are given a maintenance dose of ascorbic acid *per os* (0·5 mg per animal per day). Control animals are kept on the same diet, but have a substantially higher dose of ascorbic acid (usually 10 mg per animal per day). The body weight curves in the controls and the vitamin C deficient animals are about the same (Fig. 1). So are also the appearance, behaviour and food consumption. Figure 2 shows changes in vitamin C levels in the spleen during the development of latent vitamin C deficiency.

During the first two weeks of ascorbic acid-free feeding, vitamin C concentration falls rapidly. In the ensuing weeks, when the animals are on a maintenance dose of ascorbic acid, the level of vitamin C persists approximately at the same low values, and this regardless of the duration of the latent vitamin C deficiency. This model thus assures an equilibrated situation characterised by a stable low level of ascorbic acid in the tissues, close to concentrations that may be

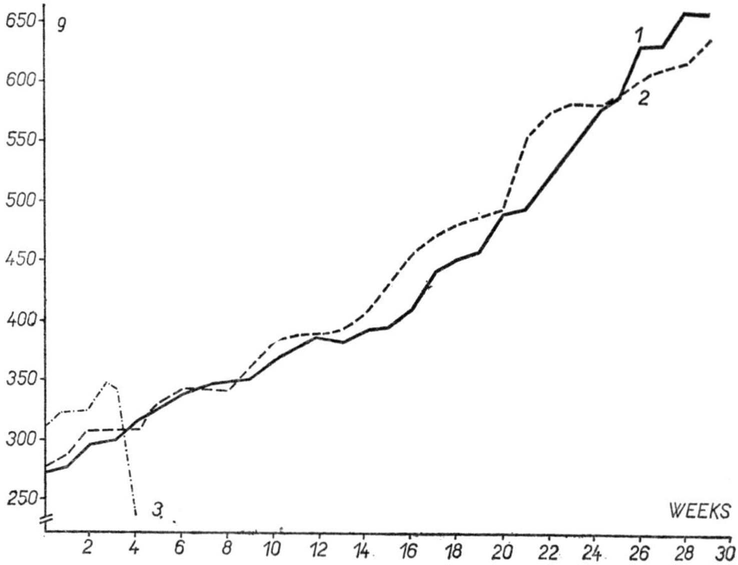

FIG. 1. Weight curves of control guinea-pigs (solid line 1), of guinea-pigs with chronic hypovitaminosis C (broken line 2) and with acute scurvy (dot-and-dashed line 3).

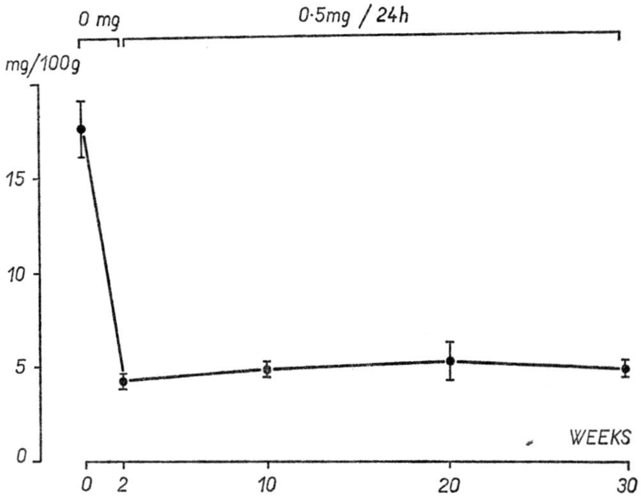

FIG. 2. Ascorbic acid concentration in the spleen during the development of chronic hypovitaminosis C in guinea-pigs. The vertical bars represent the standard errors of the mean.

observed in animals with incipient scurvy. On the other hand, a permanent supply of maintenance doses of ascorbic acid prevents the onset of acute scurvy, so that this state, similar to the one in subclinical vitamin C deficiency in humans, may be termed hypovitaminosis C. An important feature in this model is that when evaluating hypovitaminosis C, the only variable that has to be taken into account is the substantial depletion of body pool in ascorbic acid, while various secondary phenomena associated with acute scurvy (*e.g.* loss of body weight) are here immaterial. Thus, in the case of a biochemical disorder in vitamin C-deficient guinea-pigs, a decline of vitamin C concentration in blood and tissues may be unequivocally cited as the causal factor.

THE ROLE OF ASCORBIC ACID IN CHOLESTEROL METABOLISM

No significant changes of cholesterol levels become apparent in the blood and tissues of guinea-pigs during hypovitaminosis C that persists for a shorter period of time. However, in protracted hypovitaminosis C (lasting over 10 weeks) there is regularly a considerable cholesterol accumulation in the liver.[33–36] The occurrence of hypercholesterolaemia is frequent, though it sometimes fails to achieve statistical significance with smaller groups because of considerable individual variations. Table 2 shows the concentrations of total cholesterol in blood serum and in a number of tissues from a large sample group of guinea-pigs kept in a state of hypovitaminosis C for a period of 3–5 months.[37]

In the deficient group, cholesterol concentration in serum and liver is significantly increased, and the quantity of Liebermann–Burchard positive sterols in the skin also rises significantly. Cholesterol concentrations in all the other organs remained unchanged. If the guinea-pigs are fed an atherogenic diet with the addition of 0·3% cholesterol, hypovitaminosis C causes higher cholesterol levels also in further organs.[29,38] Table 3 shows the results of two experiments, lasting 12 and 20 weeks respectively, in which the effect of three graded doses of ascorbic acid (0·5–5—50 mg per animal per day) on cholesterol levels in the tissues of guinea-pigs fed an atherogenic diet, were followed.

The cholesterol levels varied in dependence on the dose of ascorbic

TABLE 2

Total cholesterol concentrations in control guinea-pigs and guinea-pigs with chronic hypovitaminosis C (mg per 100 *g wet tissue or* 100 *ml serum). In the table means from* 25 *to* 26 *animals* ± *S.E.M. are given.*

Tissue	Control	Hypovitaminosis C
Blood serum	126 ± 9	218 ± 17[a]
Liver	359 ± 15	443 ± 19[a]
Kidney	325 ± 6	325 ± 6
Adrenal	7 269 ± 384	6 844 ± 461
Small intestine	238 ± 7	231 ± 5
Large intestine	265 ± 6	268 ± 7
Stomach	329 ± 7	331 ± 6
Lung	401 ± 8	412 ± 6
Heart muscle	163 ± 4	169 ± 3
Brain	1 659 ± 32	1 658 ± 37
Testis	228 ± 7	233 ± 7
Epididymal fat	111 ± 5	122 ± 35
Skeletal muscle	86 ± 3	91 ± 3
Thoracic aorta	392 ± 46	405 ± 42
Skin	171 ± 6	211 ± 8[a]

[a] The difference is statistically significant ($P < 0{\cdot}002$–$0{\cdot}001$).

acid, the highest being found in the hypovitaminosis group (0·5 mg vitamin C per 24 hr). An enhanced cholesterol accumulation was observed in the thoracic aorta of these animals and pathomorphological changes of their blood vessels were also most pronounced.[38] Correlation analysis showed that a decline of ascorbic acid concentration in the tissues was accompanied by a significant rise in cholesterol level, and vice versa.

Cholesterol concentration in blood and tissues is the resultant of a whole series of interrelated processes, such as cholesterol distribution between blood and tissues, exogenous cholesterol absorption, biosynthesis of endogenous cholesterol, cholesterol secretion into the bile and the gastro-intestinal tract, and rate of cholesterol transformation into bile acids. In an unpublished study we followed the distribution of labelled cholesterol between the blood serum and 14 tissues as shown in Table 2. The results obtained from controls and animals with hypovitaminosis C proved to be very similar; only the passage of cholesterol from blood to skin and brain is perhaps somewhat faster in the latter. In no case, however, may the raised cholesterol levels in blood serum and liver be ascribed to a lowered rate of

TABLE 3

Total cholesterol concentrations in the tissues of cholesterol-fed guinea-pigs given various doses of ascorbic acid. In the table means from 9 *to* 14 *animals* ± *S.E.M. are given.*

Duration of experiment	Tissue	Doses of ascorbic acid (mg/24 hr/animal)		
		0·5	5	50
12 weeks	Liver	4 017 ± 485	3 652 ± 310	3 404 ± 42
	Adrenal	10 774 ± 1 621	10 647 ± 1 047	8 651 ± 527
	Small intestine	387 ± 19	345 ± 35	272 ± 32
	Thoracic aorta	548 ± 48	545 ± 96	409 ± 29
20 weeks	Liver	6 622 ± 548	5 611 ± 416	3 509 ± 350
	Adrenal	7 942 ± 890	7 782 ± 671	5 186 ± 840
	Small intestine	364 ± 23	317 ± 17	282 ± 20

cholesterol deposition in the other parts of the body of deficient animals, nor to an enhanced cholesterol absorption from the diet. As a matter of fact, ^{14}C distribution after an intragastral administration of 4-^{14}C-cholesterol has shown that absorption of exogenous cholesterol in hypovitaminosis C is actually depressed.[39] The rate of 1-^{14}C-acetate incorporation into hepatal cholesterol in hypovitaminosis C, in contrast to that in acute scurvy[8,9] is unchanged or moderately lowered.[34,36] Evidently, the function of the tricarboxylic acids cycle remains unimpaired in hypovitaminosis C, for the oxidation of 1-^{14}C-acetate to $^{14}CO_2$ proceeds normally in these animals.[36] Nor can the increased cholesterol levels be explained on the grounds of a lowered cholesterol excretion into the gastro-intestinal tract, for the output of ^{14}C-neutral sterols through the stool following an intraperitoneal administration of 4-^{14}C-cholesterol, is not affected by hypovitaminosis C.[40]

After an intraperitoneal injection of 4-^{14}C-cholesterol to hypovitaminous guinea-pigs we observed a lowered ^{14}C-bile acids output

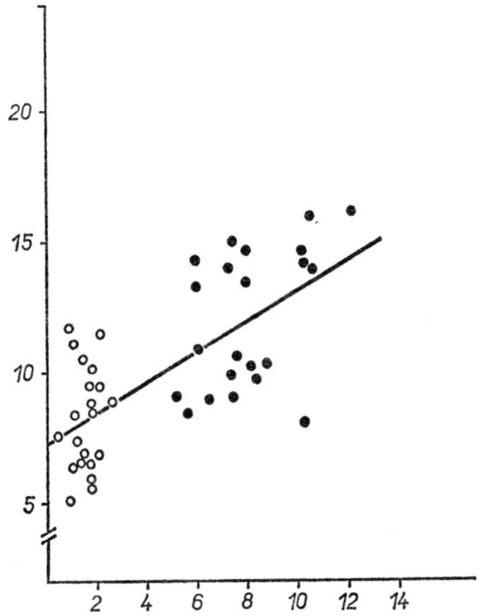

FIG. 3. Linear correlation between liver ascorbic acid concentration in mg/100 g (abscissa) and the rate of cholesterol transformation to bile acids in mg/day/500 g body weight (ordinate) in control guinea-pigs (●) and guinea-pigs with latent hypovitaminosis C (○). Statistical significance: $P < 0{\cdot}001$.

through the stool, a slowed down transfer of the label from 4-^{14}C-cholesterol to bile acids in the liver and a lowered 26-^{14}C-cholesterol oxidation to $^{14}CO_2$.[40] These results imply that the cholesterol transformation to bile acids is slowed down in hypovitaminosis C. A simultaneous determination of the quantity of expired $^{14}CO_2$ and of the specific activity of hepatal or serum cholesterol after administration of 26-^{14}C-cholesterol, makes it possible for the rate of cholesterol transformation to bile acids to be quantified.[41,42] Making use of this procedure, we have succeeded in showing that a latent vitamin C deficiency significantly slows down the rate of cholesterol transformation to bile acids[37,43] (controls: $11{\cdot}8 \pm 0{\cdot}6$; hypovitaminosis C: $8{\cdot}3 \pm 0{\cdot}4$ mg of cholesterol per 24 hr per 500 g body weight; $P < 0{\cdot}001$). Cholesterol is transformed to bile acids in the liver, and the rate of this process seems to depend on the vitamin C concentration in the liver cells, as there is a linear correlation between the rate of cholesterol transformation to bile acids and vitamin C concentration in the liver (Fig. 3).

The decrease in the concentration of hepatal vitamin C causes a decrease in the rate of cholesterol conversion to its main catabolic product, the bile acids, with the result that cholesterol accumulates in the blood and certain tissues. Conversely, resaturation of vitamin C deficient guinea-pigs with high doses of ascorbic acid strikingly stimulates 26-^{14}C-cholesterol oxidation to $^{14}CO_2$ (Fig. 4).[44]

The conversion of cholesterol to bile acids is a multistage process which takes place in the microsomes, supernatant fraction and mitochondria of the liver cell, successively, and comprises hydroxylation, dehydrogenation, saturation of the nuclear double bond, reduction of 3-ketone and ω- and β-oxidation of the side chain.[45,46] It is difficult to decide by what mechanism ascorbic acid participates in the catabolism of cholesterol. By analogy with the general function of ascorbic acid in hydroxylation reactions,[1,2,47–51] we suggested, as a working hypothesis, that ascorbic acid is necessary for the hydroxylation of cholesterol.[40] Kritchevsky *et al.*[52] have recently found that with increasing concentrations of ascorbic acid added to hepatal microsomes of normal guinea-pigs, 7α-hydroxylation of 1,2-^{3}H-cholesterol gradually increases, but the results fail to be of statistical significance. Our results indicate that the effect of ascorbic acid on cholesterol catabolism might be indirect[53] mediated by its action on the cytochrome P-450 level in the liver cell microsomes.[54] It is likewise possible that ascorbic acid stimulates the formation of cholesterol sulphate which is water-soluble and so facilitates cholesterol clearance.[55]

Table 4 presents the results of a kinetic two-pool analysis[25] of the die-away curves of plasma cholesterol specific activity after one pulse labelling with cholesterol-4-^{14}C.

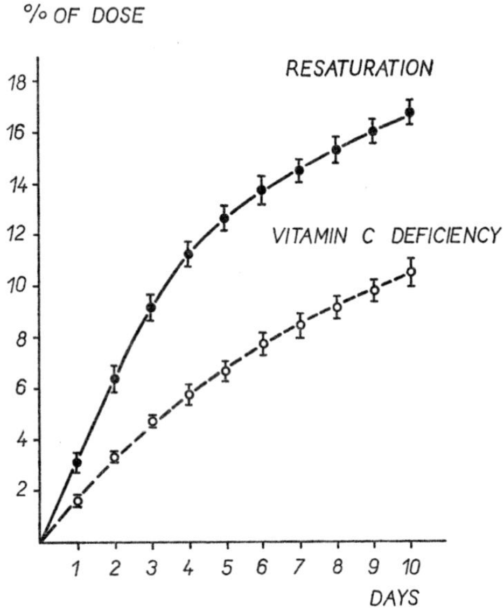

FIG. 4. Oxidation of 26-^{14}C-cholesterol to $^{14}CO_2$ as percentage of the dose injected in vitamin C deficient and resaturated guinea-pigs. The differences from the second day are statistically highly significant. The vertical bars represent the standard errors of the mean.

In guinea-pigs with hypovitaminosis C, the half-life of the linear part of the die-away curve was significantly prolonged, the value of the rate constant for the irreversible excretion of cholesterol and cholesterol turnover rate significantly decreased. Hence, a latent vitamin C deficiency induced in guinea-pigs changes similar to those we described in rabbits with atherosclerosis provoked by a cholesterol rich diet (Table 1): increased cholesterol concentration in blood plasma, delayed cholesterol release from the circulation and from the entire organism. In this manner, hypovitaminosis C creates a metabolic situation in which the risk of an atheromatous rebuilding of the blood vessels is increased.

In guinea-pigs kept in a state of hypovitaminosis C for over 6 months, we found oedema of the vascular wall being formed in their aorta, vacuolisation of the endothelium and a parietal adhesion of blood plasma. In the intima of the arteries and coronary branches in part of the hypovitaminous animals the formation of foam-cells,

TABLE 4

The influence of chronic hypovitaminosis C on the size of pools and kinetic parameters of cholesterol turnover in guinea-pigs. In the table means from 9 to 12 animals ± S.E.M. are given

	Parameter	Control	Hypo-vitaminosis C
$t_{1/2\alpha}$	Half-life of first exponential (days)	8·0 ± 1·3	7·1 ± 2·0
$t_{1/2\beta}$	Half-life of second exponential (days)	24·0 ± 1·1	30·1 ± 1·7
M_A	Size of pool A (mg)	1 084 ± 78	1 071 ± 79
$M_{B\,min}$	Minimum size of pool B (mg)	160 ± 39	258 ± 73
PR_A	Production rate in pool A = turnover rate (mg per 24 hr)	37·8 ± 2·0	31·8 ± 1·5
k_A	Rate constant for irreversible excretion from pool A (day^{-1})	0·036 ± 0·001	0·031 ± 0·002

In guinea-pigs with chronic hypovitaminosis C half-life of second exponential is prolonged ($P < 0{\cdot}01$); turnover rate and rate constant for irreversible excretion of cholesterol are significantly decreased ($P < 0{\cdot}05$).

and even fresh homogenised atheromatous masses were observed (Fig. 5).

It should be noted that these findings come from animals fed a diet without the addition of cholesterol, so that the presclerotic, even atheromatous changes can be ascribed exclusively to a long-term latent deficiency of vitamin C. In hypovitaminous guinea-pigs we regularly noted also an increased triglyceride concentration in blood plasma. The mechanism of this phenomenon is unexplained, but it is probably related to the action of ascorbic acid on lipase activity in the blood and tissues.[19,56,57] Besides disorders of lipid metabolism, disorders in the metabolism of the connective tissue in the vascular wall[27,28] and of the anticoagulating system in the blood[1,58] of vitamin C-deficient animals, might also participate in the onset of atheromatous changes.

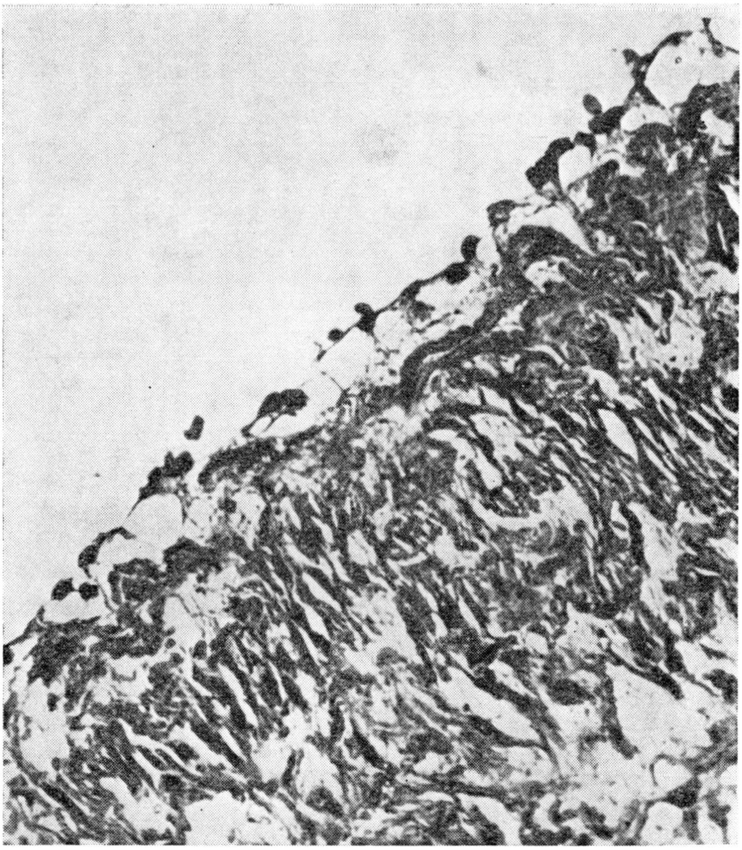

FIG. 5. Early atheromatous lesions in the intima of thoracic aorta of hypovitaminotic guinea-pig fed basal diet without addition of cholesterol. Duration of latent hypovitaminosis C: 28 weeks.

VITAMIN C, HYPERCHOLESTEROLAEMIA AND ATHEROSCLEROSIS IN HUMANS

Among the numerous papers about the effect of ascorbic acid on cholesterolaemia in men there is no agreement concerning the influence of vitamin C on the level of cholesterol in blood.[2] With the exception of a single paper[59] all the authors agree that the administration of various doses of ascorbic acid fails to affect serum cholesterol levels in persons with normal cholesterol values.[60-65] Several authors of clinical studies[19,60,61,66,67] performed on men with hypercholesterolaemia and atherosclerosis report that under the effect of vitamin C hypercholesterolaemia recedes in at least part of the patients, with a decline in β-lipoproteins, resulting in an improved health condition. Most of these papers are criticised[2] on the ground that the hypocholesterolaemic effect has been attained under simultaneous influence of further factors such as protective diet, curative regime, etc. Reports are also available according to which vitamin C has no effect on plasma cholesterol levels even in patients with hypercholesterolaemia and atherosclerosis,[68,69] or it even increases plasma cholesterol levels.[59] The assumption according to which this increase is induced by a mobilisation of arterial cholesterol, sounds very improbable.[70] Serum cholesterol level in scorbutic patients remains unchanged,[71] or becomes increased.[64,72] Following treatment with ascorbic acid, it generally rises.[64,72] The mechanism of this phenomenon is not clear. Its interpretation is made the more difficult, just as in the case of vitamin C deficient animals, by a complex metabolic disorganisation provoked by scurvy. Various authors pointed out the seasonal fluctuation of cholesterolaemia in men:[73–75] the highest concentrations of serum cholesterol could be observed in the period of minimal intake of vitamin C. But, whether a causal relation is involved, remains as yet an open issue. Similarly, an attempt at correlating the levels of ascorbic acid and cholesterol in a single blood sample[76,77] will yield little information, for the actual concentration of vitamin C in the blood provides no information on the nutritional history of the persons investigated. Despite this, several authors[78–80] have described a statistically significant negative correlation between vitamin C status and cholesterolaemia. Knox[81] carried out correlation analysis between the standardised mortality ratios for ischaemic heart disease and cerebrovascular disease in different regions of England in different years and dietary intakes of a number of nutrients. Vitamin C intakes showed a strong negative correlation: in the regions with low vitamin C intake the mortality for ischaemic heart disease and cerebrovascular disease was high and vice versa.

If the results of our experiments summarised in the previous section could be applied to the human organism, it would mean that a long-term latent vitamin C deficiency induces also in men a slowed down cholesterol transformation to bile acids, resulting in hypercholesterolaemia and a delayed release of cholesterol from the circulation. Administration of high doses of vitamin C should, similarly as in guinea-pigs,[44] enhance cholesterol catabolism, and after a prolonged period, lead to a decline of cholesterolaemia. To test this the effect of vitamin C on cholesterolaemia in a selected group of persons above 40 years of age with a seasonal deficit of ascorbic acid and with an initial level of serum cholesterol in the upper permitted limit, or hypercholesterolaemia, was studied.[82] Administration of 300 mg ascorbic acid daily for 47 days depressed cholesterolaemia moderately, but to a statistically significant degree. The effect of vitamin C was most pronounced in persons with high starting cholesterol levels (Table 5).

Hypercholesterolaemia may be induced by a number of factors. In cases where it has been produced by a factor other than vitamin C deficiency, it is improbable that administration of ascorbic acid will be effective. Hence, in our view, it is superfluous to administer ascorbic acid to hypercholesterolemic patients well supplied with vitamin C. Our conception of the relationship between vitamin C and

TABLE 5

The influence of ascorbic acid (300 *mg*/24 *hr*) *on serum total cholesterol concentrations* (*mg*/100 *ml*) *in hypercholesterolaemic persons with seasonal vitamin C deficiency*

Person examined	Serum cholesterol (mg/100 ml)		Difference
	January 1968	March 1968	
K.M. ♀	340	260	−80
K.M. ♀	320	260	−60
R.A. ♀	313	321	+ 8
C.A. ♂	305	275	−30
M.J. ♂	305	248	−57
P.J. ♂	291	199	−92
K.A.♂	291	260	−31
O.J. ♂	291	283	− 8
B.L. ♂	269	283	+14
M.H. ♀	255	217	−38
F.F. ♂	254	220	−34
S.A. ♂	247	231	−16
P.J. ♂	243	239	− 4
Mean ± S.E.M.	286 ± 9	253 ± 9	−33 ± 9
Statistical significance:			$P < 0{\cdot}01$

cholesterol turnover emphasises the *preventive aspect* of the problem: care should be taken to prevent a state of latent ascorbic acid deficiency and thereby forestall any disorder in the process of cholesterol transformation to bile acids. We have shown that such a disorder in experimental animals leads to hypercholesterolaemia and atherosclerosis and this may possibly apply also to the human organism. The fact that also in countries with a high living standard, the intake of vitamin C in certain groups of the population is very low,[83-87] only goes to underline the urgency of this problem. We are of the opinion that a long-term latent vitamin C deficiency in people should be considered as a factor enhancing the risk of atherogenesis.

CONCLUSION

In an extensive series of experiments on animals we have shown that latent vitamin C deficiency provokes a metabolic disorder in the liver, causing an impaired cholesterol transformation to its principal catabolic products, bile acids. This metabolic disorder induces hypercholesterolaemia and slows down the release of cholesterol from the circulation. When such a metabolic state persists over a long period, an atheromatous rebuilding of the vascular system may ensue in a vitamin C deficient organism. It is thought probable that a latent vitamin C deficiency has similar consequences also for the human organism. The data obtained assign the latent deficiency of ascorbic acid to the group of risk factors in the pathogenesis of atherosclerosis and emphasise the importance of prevention of hypovitaminosis C.

ACKNOWLEDGEMENTS

I thank J. Babala, M.D., C.Sc. (Komensky University, Bratislava) for histological research, Ing. R. Nemec, Ing. J. Cerven and L. Mikus (Isotope Laboratory of the Institute of Human Nutrition Research, Bratislava) for cooperation in the studies with labelled cholesterol.

REFERENCES

1. Sebrell, W. H. and Harris, R. S. (1967). *The Vitamins*, Vol. I., Academic Press, New York and London.

2. Ginter, E. (1970). *The Role of Ascorbic Acid in Cholesterol Metabolism*, Ed. SAV, Bratislava.
3. Banerjee, S. and Bandyopadhyay, A. (1963). *Proc. Soc. exp. Biol. (N.Y.)*, **112,** p. 372.
4. Ginter, E., Bobek, P. and Gerbelová, M. (1965). *Nutr. Dieta,* **7,** p. 103.
5. Fujinami, T., Okado, K., Senda, K., Sugimura, M. and Kishikawa, M. (1971). *Japanese Circul. J.*, **35,** p. 1559.
6. Banerjee, S. and Singh, H. D. (1958). *J. biol. Chem.*, **233,** p. 336.
7. Kawishwar, W. K., Chakrapani, B. and Banerjee, S. (1963). *Ind. J. med. Res.*, **51,** p. 488.
8. Becker, R. R., Burch, H. B., Salomon, L. L., Venkitasubramanian, T. A. and King, C. G. (1953). *J. Amer. Chem. Soc.*, **75,** p. 2020.
9. Banerjee, S. and Ghosh, P. K. (1960). *Amer. J. Physiol.*, **199,** p. 1064.
10. Guchhait, R., Guha, B. C. and Ganguli, N. C. (1963). *Biochem. J.*, **86,** p. 193.
11. Hodges, J. R. and Hotston, R. T. (1970). *Brit. J. Pharmacol.*, **40,** p. 740.
12. Kitabchi, A. E. and Duckworth, W. C. (1970). *Amer. J. Clin. Nutr.*, **23,** p. 1012.
13. Chevallier, F. (1967). *Adv. Lipid. Res.*, **5,** p. 209.
14. Mueller, P. S. (1962). *J. Lipid Res.*, **3,** p. 92.
15. Abramson, H. (1949). *J. biol. Chem.*, **178,** p. 179.
16. Takeda, Y. and Hara, M. (1955). *J. biol. Chem.*, **214,** p. 657.
17. Mjasnikov, A. L. (1958). *Circulation*, **17,** p. 99.
18. Zajcev, V. F., Mjasnikov, L. A., Kasatkina, L. V., Lobova, N. M. and Sukasova, T. I. (1964). *Cor Vasa*, **6,** p. 18.
19. Sokoloff, B., Hori, M., Saelhof, C., McConnel, B. and Imai, T. (1967). *J. Nutrition,* **91,** p. 107.
20. Flexner, J., Bruger, M. and Wright, I. S. (1941). *Arch. Pathol.*, **31,** p. 82.
21. Volkova, K. G. (1961). *Ateroskleroz, Medgiz, Leningrad,* p. 187.
22. Pool, W. R., Newmark, H. L., Dalton, C., Banziger, R. F. and Howard, A. N. (1971). *Atherosclerosis*, **14,** p. 131.
23. Rolek, D. F. and Dale, H. E. (1927). *Atherosclerosis*, **15,** p. 185.
24. Ginter, E., Babala, J. and Polónyová, E. (1970). *Biológia (Bratislava)*, **25,** p. 579.
25. Nestel, P. J., Whyte, H. M. and Goodman, DeW. S. (1969). *J. clin. Invest.*, **48,** p. 982.
26. Willis, G. C. (1957). *Canad. med. Ass. J.*, **77,** p. 106.
27. Gore, I., Fujinami, T. and Shirahama, T. (1965). *Arch. Pathol.*, **80,** p. 371.
28. Gore, I., Tanaka, Y., Fujinami, T. and Goodman, M. L. (1965). *J. Nutrition*, **87,** p. 311.
29. Ginter, E., Bobek, P., Babala, J. and Barbieriková, E. (1969). *Cor Vasa*, **11,** p. 65.
30. Ginter, E., Bobek, P. and Ovecka, M. (1968). *Inter. J. Vitamin Res.*, **38,** p. 104.

31. Babala, J. and Ginter, E. (1968). *Nutr. Dieta*, **10,** p. 133.
32. Ginter, E., Babala, J., Bobek, P. and Dumbalová, Z. (1968). *Cor Vasa*, **10,** p. 126.
33. Ginter, E., Ondreicka, R., Bobek, P. and Simko, V. (1969). *J. Nutrition*, **99,** p. 261.
34. Ginter E., Bilisics, L. and Cerven, J. (1965). *Physiol. bohemoslov*, **14,** p. 466.
35. Ginter, E., Bobek, P., Kopec, Z., Ovecka, M. and Cerey, K. (1967). *Z. Versuchstierk.*, **9,** p. 228.
36. Ginter, E. and Nemec, R. (1969). *J. Atheroscler. Res.*, **10,** p. 273.
37. Ginter, E., Nemec, R., Cerven, J. and Mikus, L. *Lipids*, **8,** p. 135.
38. Ginter, E., Babala, J. and Cerven, J. (1969). *J. Atheroscler. Res.*, **10,** p. 341.
39. Ginter, E., Cerven, J. and Mikus, L. (1969). *Physiol. bohemoslov.*, **18,** p. 459.
40. Ginter, E., Cerven, J., Nemec, R. and Mikus, L. (1971). *Am. J. clin. Nutr.*, **24,** p. 1238.
41. Myant, N. B. and Lewis, B. (1966). *Clin. Sci.*, **30,** p. 117.
42. Ginter, E., Nemec, R., Cerven, J., Bobek, P., Mikus, L. and Jankela, J. (1973). *Physiol. bohemoslov.*, **22,** p. 287.
43. Ginter, E. (1973). *Science*, **179,** p. 702.
44. Ginter, E., Nemec, R. and Bobek, P. (1972). *Brit. J. Nutr.*, **28,** p. 205.
45. Elliott, W. and Hyde, P. M. (1971). *Am. J. Med.*, **51,** p. 568.
46. Boyd, G. S. and Percy-Robb, I. W. (1971). *Am. J. Med.*, **51,** p. 580.
47. Degkwitz, E. and Staudinger, Hj. (1965). *Hoppe-Seyler's Z. Physiol. Chem.*, **342,** p. 63.
48. Nakashima, Y., Suzue, R., Sanada, H. and Kawada, S. (1970). *J. Vitaminol.*, **16,** p. 276.
49. Barnes, M. J. and Kodicek, E. (1972). *Vit. Hormones*, **30,** p. 1.
50. Wagstaff, D. J. and Street, J. C. (1971). *Toxicol. appl. Pharmacol.*, **19,** p. 10.
51. Zannoni, V. G., Flynn, E. J. and Lynch, M. (1972). *Biochem. Pharmacol.*, **21,** p. 1377.
52. Kritchevsky, D., Tepper, S. A. and Story, J. A. (1973). *Lipids*, **8,** p. 482.
53. Ginter, E. and Nemec, R. (1972). *Physiol. bohemoslov.*, **21,** p. 539.
54. Leber, H. W., Degkwitz, E. and Staudinger, Hj. (1970). *Hoppe-Seyler's Z. Physiol. Chem.*, **351,** p. 995.
55. Mumma, R. O. and Verlangieri, A. J. (1971). *Fed. Proc.*, **30,** p. 370.
56. Kotzé, J. P., Matthews, M. J. A. and De Klerk, W. A. (1973). *Lancet*, **1,** p. 610.
57. Milenkov, H. and Mitkov, D. (1969). *Folia med.*, **11,** p. 191.
58. Andrejenko, G. V. and Ljutova, L. V. (1971). *Bull. eksp. biol. med.*, **71**(6), p. 31.
59. Spittle, C. R. (1971). *Lancet*. **11,** p. 1280.
60. Tjapina, L. A. (1952). *Gipertoniceskaja bolezn. Trudy AMN SSSR*, **2,** p. 108.
61. Bukovskaja, A. V. (1957). *Sovetsk. med.*, **21**(1), p. 77.
62. Anderson, J., Grande, F. and Keys, A. (1958). *Fed. Proc.*, **17,** p. 468.

63. Hrubá, F. and Masek, J. (1962). *Nahrung*, **6,** p. 507.
64. Bronte-Stewart, B., Roberts, B. and Wells, V. M. (1963). *Brit. J. Nutr.*, **17,** p. 61.
65. Anderson, T. W., Reid, D. B. W. and Beaton, G. H. (1972). *Lancet*, **11,** p. 876.
66. Sedov, K. R. (1956). *Terap. Arch.*, **28**(2), p. 58.
67. Fedorova, E. P. (1960). *Sovetsk. Med.*, **21**(11), p. 56.
68. Samuel, P. and Salchi, O. B. (1964). *Circulation*, **29,** p. 24.
69. Krivorucenko, I. V. (1963). *Terap. Arch.*, **35**(4), p. 48.
70. Morin, R. J. (1972). *Lancet*, **1**, p. 594.
71. Hodges, R. E., Baker, E. M., Hood, J., Sauberlich, H. E. and March, S. C. (1969). *Am. J. clin. Nutr.*, **22,** p. 535.
72. Hodges, R. E., Hood, J., Canham, J. E., Sauberlich, H. E. and Baker, E. M. (1971). *Am. J. clin. Nutr.*, **24,** p. 432.
73. Thomas, C. B., Holljes, H. W. D. and Eisenberg, F. F. (1961). *Ann. intern. Med.*, **54,** p. 413.
74. Tochowicz, L., Ciba, T., Kocemba, J. and Szopinska-Ciba, L. (1962). *Polski Tyg. lek.*, **17,** p. 587.
75. Fyfe, T., Dunningan, M. G., Hamilton, E. and Rae, R. J. (1968). *J. Atheroscler. Res.*, **8,** p. 591.
76. Bradley, D. W., Maynard, J. E. and Emery, G. E. (1973). *Lancet*, **11,** p. 201.
77. Elwood, P. C., Hughes, R. E. and Hurley, R. J. (1970). *Lancet*, **11,** p. 1197.
78. Masek, J., Krikava, L. and Novák, M. (1958). *Cas. Lék. cesk.*, **97,** p. 431.
79. Masek, J. (1960). *Nutr. Dieta*, **2,** p. 193.
80. Cheraskin, E. and Ringsdorf, W. M. (1968). *Intern. J. Vit. Res.*, **38,** p. 415.
81. Knox, E. C. (1973). *Lancet*, **1,** p. 1465.
82. Ginter, E., Kajaba, I. and Nizner, O. (1970). *Nutr. Metabol.*, **12,** p. 76.
83. Hejda, S., Osancová, K. and Cibochová, E. (1965). *Cs. Gastroent. Vyz.*, **19,** p. 438.
84. Hanck, A. B., Birke, E., Brubacher, G. B., Henze, H., Müller-Mulot, W., Ohlig, W., Schulze-Falk, K. H., Sehm, G. and Vuilleumier, J. P. (1971). *Diagnostik*, **4,** p. 263.
85. Wilson, C. W. M. and Nolan, C. (1970). *Intern. J. Med. Sc.*, **3,** p. 345.
86. Silink, S. J., Nobile, S. and Woodhill, J. M. (1972). *J. Geriatrics*, **3,** p. 27.
87. Nobile, S. and Woodhill, J. M. (1973). *Food Technol.* (*Australia*), **25,** p. 1.

DISCUSSION

Kotze: We have carried out similar experiments to those of Dr. Ginter and have obtained the same sort of results, namely that a decrease of cholesterol synthesis is caused by hypovitaminosis C, and that vitamin C increases cholesterol side-chain cleavage.

Wilson: We have been carrying out work in this field also, using a different method to Dr. Ginter in reducing the ascorbic acid levels. We have found in exactly the same way that when you reduce this level you get a rise in the tissue cholesterol and the blood cholesterol, and so I would like to confirm and endorse Dr. Ginter's findings here. I think it's pretty definite that ascorbic acid does play a big part in limiting cholesterol levels and thus preventing the production of atherosclerosis.

Degkwitz: Dr. Ginter's data fit ours, for if he feeds the animals for about 14 days without ascorbic acid, the cytochrome P-450 drops down to about 50% of the starting value, and if he gives them a bit of ascorbic again it might go up to 60%. So we find a reduction of 30% in cholesterol breakdown would quite well fit our figures. If we give ascorbic acid to deficient animals over a period of 24 hr, the amounts of cytochrome P-450 return to normal; this fits his figures too.

Chairman: May I ask Professor Wilson if he has measured esterified cholesterol and free cholesterol, and whether there is any change in the ratio of the two? Also is this drop absolute or proportional?

Wilson: We were only measuring free cholesterol.

Pollitt: Shittle found that dosing with vitamin C had the effect of lowering plasma cholesterol in a group of young people, but of raising levels in an older group with a history of atherosclerosis.

When Dr. Ginter gave his figures for man he made the point, I think, that they were statistically significant but that these falls in plasma cholesterol were not by any means clinically significant. From the clinical point of view these falls were only 15 mg or 20 mg, and don't mean very much, but biochemically perhaps they do. This may just be a clue.

Wilson: I think in relation to that what one must remember is that alterations of the cholesterol levels are not absolutely related to the ascorbic acid levels—in other words, a change in ascorbic acid will alter the cholesterol over a limited range, and then presumably the metabolic balance of the body alters again, and you find a change in the triglycerides is occurring and the cholesterol change seems to cease. I don't know what the relationship is between the two—we have been analysing this, and have found there is a relationship between triglycerides, cholesterol, and ascorbic acid, and apparently they are interdependent. But where one change alters the other two we aren't quite sure yet.

Kotze: We have done some experiments with primates that were put under stress, that means wild primates were put into cages, and in this so-called induced stress their serum cholesterol went up to 70% beyond normal if they did not receive extra vitamin C during this period. If, however, we start giving them oral doses of vitamin C at the level of 2 mg per kg of body weight per day, or 5, or 10, we could keep it constant or even lower it over the initial period of stress.

Chairman: Does ascorbic acid affect the diurnal variation in cholesterol? The levels go up and down, don't they?

Wilson: We have not looked at this directly in relation to cholesterol, but there is a diurnal rhythm in ascorbic acid level.

Chairman: When is the maximum point in level of cholesterol, do you know?
Kotze: In rats it would be during the night.
Wilson: It would be the reverse in man. The peak in man would be between 8 o'clock and 12 o'clock in the morning—the maximum around 10 o'clock. It follows the cortisone level.
Kotze: We have done experiments also with vitamin C and the enzyme called lipoprotein lipase, and we have found that in the heart muscle of baboons that did not receive their ration of vitamin C, although they were not scorbutic, the level of the enzyme in the heart muscle was increased more than 200% above the levels of animals that receive vitamin C. From our studies it appeared that if the blood level had reached the serum level of vitamin C lower than 0·35 mg, then the enzyme was repressed, and if the blood level went lower than that, it started to increase rapidly.
Hornig: We have done studies with guinea-pigs involving very high amounts of ascorbic acid, and have looked at the bile acids, and the results we get seem to confirm Dr. Ginter's findings since in these guinea-pigs the bile acids were significantly enhanced in the small intestine. Another point that confirms Dr. Ginter's observation is that when we give orally 4-^{14}C labelled cholesterol to guinea-pigs maintained on 600 mg ascorbic acid per day, liver, kidney, lung and spleen will accumulate less radioactivity, *i.e.* less cholesterol, than animals which are on a normal ration of vitamin C.
Lewin: Has Professor Wilson followed the diurnal variation of cholesterol with the diurnal variation of cyclic AMP? I ask, because there is a relationship.
Wilson: I'm sure there is a relationship, but I haven't done it. I'm sure you'd like to tell us about it.
Lewin: No, I just wanted to ask if you had done the experiment.
Hammond: A lot of people have always considered fruit and vegetables as the primary source of vitamin C. Can anyone indicate the level of vitamin C in the muscles of animals and whether any particular one, such as the rabbit, is important in this respect?
Ranken: Mr. Hammond asks almost the question I wanted to ask. Am I right to take it from what the clinical investigators have been talking about today that when you refer to tissues you do refer to organs such as liver, spleen, etc., not to muscles, and that no one can quote any good figures for ascorbic acid in the muscles of animals? Can anyone point to species differences among animals, and in particular Mr. Hammond's point, does the rabbit have higher levels of ascorbic acid in its muscles than other species?
Wilson: The answer is you have to differentiate between human beings and guinea-pigs (which a lot of the discussion has been about today) and animals which can synthesise their own vitamin C. In guinea-pigs and human beings the level varies. In the skin of human beings it's about 0·4 mg per cent and it goes up to a level of 1½ mg per cent in the liver and heart, and in a human receiving the maximum doses of vitamin C (10 g a day) over a long period, the maximum level recorded is 2½ mg per cent in the

plasma. Now guinea-pig levels are approximately the same. Animals such as the rabbit are more or less saturated all the time and therefore their levels would be somewhere in the region of 1–2 mg per cent (uncooked).
Ranken: This agrees with the textbook statements for the ascorbic acid content of meats (from pigs and cattle and so forth)—levels generally called negligible in ordinary meats, but going up to levels such as 1·5 or as high as 3 in organs such as the liver and kidneys and the brain.
Hornig: May I draw your attention to an article by Kirk in *Vitamins and Hormones*, **20,** pp. 83–92, which appeared in 1962 and which gives something on ascorbic acid muscle content. The title is 'Variations in the Tissue Content of Vitamins and Hormones', Chapter IV, Ascorbic Acid.
Chairman: May I ask Dr. Hughes—he has told us something about ascorbic acid and its effect on mercury. Does he know of any work on its effect on lead in tissues? We are a great deal concerned nowadays about lead as a contaminant.
Hughes: The only reference I can think of is one where some studies have been done which showed that ascorbic acid enhanced the uptake of cadmium and lead.
Chairman: I know that there is some work going on at the moment in the Isle of Sheppey, in which the blood lead levels of children are being measured by the Great Ormond St. Hospital. It would be very interesting to know the ascorbic acid levels here.
Hughes: Yes, if one gets a considerable spread in the lead levels, possibly it would be interesting to see if there was a similar correlation with ascorbic acid.
Wilson: Can Dr. Hughes give us any information about the reduced levels of ascorbic acid in smokers' blood?
Hughes: No. We tried to repeat this with guinea-pigs some years ago—we devised a smoking machine. It produced no differences except in the adrenal levels, and we put this down to the stress of smoking—though there were no obvious signs of stress, they appeared to like it. In more recent studies in Cardiff we divided a group of elderly persons into high ascorbic acid subjects and low ones (those with a level above and below 0·2), subdivided into smokers and non-smokers, and we did find a correlation in what we called the fruit intake and smoking. In general, the people who smoked ate less fruit. The statisticians found you could almost—not completely—account for the difference in terms of the different fruit intake. There is a second factor, which is probably metabolic.
Wilson: I suggest that that might be due to the cadmium intake in smoke.
Hughes: You would have to have a fairly substantial level of cadmium to get any sort of reduction. But it could be an additional factor, certainly.
Lewin: It might interest you to know that if you install an electrostatic air filter in your home you will find it contains quite definite quantities of lead.
Goodman: I have been reading an article on work being done at the Institute of Urology correlating chemiluminescence in urine with bladder cancer and smoking, and it suggested that by giving $1\frac{1}{2}$ g of vitamin C daily for a few days they could reduce the chemiluminescence. This seems to be very much related to the incidence of bladder cancer in the chemical

industries, and to the implications of metals in tobaccos, as previous speakers suggested.

Ranken: I think the problem there is that we spoke to the biochemists who were involved in that work, and they haven't been able to identify the cause of the chemiluminescence. I don't think we can say a great deal on this at the moment, until the substance has been identified.

Rivers: It might be relevant here to comment that we can still cure scurvy because we can identify the element responsible for it.

14

Vitamin C: Tissue Metabolism, Over-saturation, Desaturation and Compensation

C. W. M. Wilson

Department of Pharmacology, University of Dublin, Trinity College, Dublin, Ireland

ABSTRACT

Vitamin C is present in all human tissues. Ascorbic acid is a labile substance which passes rapidly from the plasma into the white cells where it is involved in immunological defence, and also is stored. It passes from the plasma into the body tissues where it plays important roles in liver metabolism, haemopoiesis, collagen formation, endocrine control and metabolism including ovulation, and other metabolic processes. General tissue metabolism is maintained by ascorbic acid, and tissue anabolism and growth is stimulated by a supply of the vitamin adequate to maintain tissue saturation. Humans and guinea-pigs require exogenous vitamin C for these metabolic purposes. The four-fold range in the recommended dietary intake throughout the world is explicable in terms of the negative concept that only enough ascorbic acid is required by the tissues to prevent scorbutic signs, or by the positive concept that the vitamin can improve tissue efficiency and integrity when sufficient is available to maintain saturation. Tissue requirements vary between individuals, and are affected by dietary intake, age, sex, and the occurrence of patho-physiological states. When leucocyte stores become saturated in normal individuals, plasma concentrations increase, and renal excretion becomes maximal. Side effects are stated to follow administration of supplementary vitamin C to humans in such circumstances, including alimentary disturbances, oxalic acid stone formation, and infertility. The evidence supporting such statements is unsubstantiated. Patho-physiological states cause increased tissue utilisation of ascorbic acid, and the consequent appearance of isolated scorbutic signs and symptoms in association with localised reduction of ascorbic acid in these tissues, and/or reduction in leucocyte stores and a fall in plasma ascorbic acid. Humans and guinea-pigs respond by making readjustments in their ascorbic acid utilisation and metabolism which is particularly pronounced in females.

Ascorbic acid is present in all human tissues.[1,2] It undergoes a circadian rhythm in the plasma which is correlated with that in the leucocytes.[3] This rhythm is synchronous with the plasma iron rhythm.[4] Iron transfer from the plasma into the liver through the ferritin mechanism is dependent upon the availability and reducing power of ascorbic acid.[5] Ascorbic acid undergoes a monthly rhythm in relation to ovulation,[6] and plays important roles in haemopoiesis and liver metabolism (Fig. 1). It is involved in the synthesis of lipoproteins, plasma and tissue proteins and also in collagen formation. This last is essential for tissue healing, and is associated with an energy dependent process involving retention of ascorbic acid by the tissues. Animal species which can synthesise their own ascorbic acid (that is all except primates, guinea-pigs and the fruit-bat),

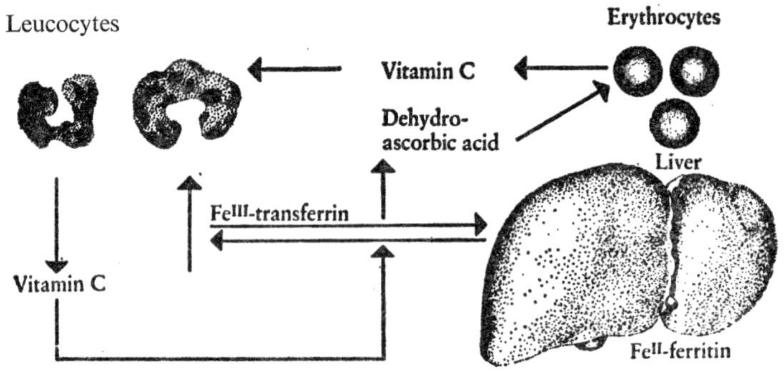

FIG. 1. Transfer of vitamin C from the erythrocytes through the plasma to the leucocytes where it is stored in a labile form. Ascorbic acid is involved in the hepatic transfer of iron through the transferrin–ferritin mechanism, when it becomes oxidised to dehydro-ascorbic acid. The dehydro-ascorbic acid is then rapidly transferred through the erythrocytes while being reduced again to ascorbic acid (Wilson[5]).

maintain a high level of tissue saturation. There is a four-fold range in the recommended human dietary intake of vitamin C. This varies from 30 mg in the UK to 125 mg in the USSR.[7] The lower intake is explicable in terms of the negative concept that a daily intake only sufficient to prevent the occurrence of scurvy, is recommended by the UK Panel on Recommended Allowances of Nutrients.[8] Higher intakes are based on the positive concept that vitamin C can improve tissue efficiency and integrity when available to human beings in sufficient quantity to maintain tissue saturation. In these circumstances, vitamin C exerts a positive role in promoting tissue anabolism and growth. Its lability enables its easy and rapid transfer so that optimal efficiency is maintained in all tissues.

Administration of vitamin C to normal human beings causes an increase in mean leucocyte ascorbic acid concentrations. In conditions of dietary deficiency or stress, desaturation of particular tissues may occur. This develops into a state of general desaturation as the deficiency becomes more widespread and tissue stores are utilised. Depletion of tissue ascorbic acid accompanies many pathological states. States of over-saturation and desaturation will be described, and tissue reaction to depletion will be discussed in circumstances when supplementary vitamin C is not administered.

THE METABOLISM OF ASCORBIC ACID

Ingested vitamin C is absorbed from the alimentary canal by a passive process which is pH dependent and is related directly to the duration and extent of contact with the absorbing surface.[9,10] Absorption is impaired during the later stages of vitamin C deficiency possibly in consequence of damage to the lining membrane of the alimentary canal, or as a result of pituitary disfunction.[11,12] In normal circumstances the vitamin C is absorbed from the alimentary canal and passes into the plasma. It is absorbed into the leucocytes by active and passive mechanisms,[13] where it is stored provided that tissue demands have been satisfied[5] and vitamin utilisation is stable.[14]

The extent to which ascorbic acid is metabolised can be evaluated by measurement of the plasma and leucocyte ascorbic acid values after a loading dose of vitamin C. Measurement of the urinary excretion of ascorbic acid during the 4-hr period following the loading dose indicates whether the body is conserving ascorbic acid under specific conditions in comparison with readministration of the dose in different circumstances. When utilisation is enhanced, it may be necessary to administer up to 2000 mg vitamin C as a loading dose in order to obtain the normal rise in leucocyte ascorbic acid.[15]

Ascorbic acid is excreted as a result of glomerular filtration and active tubular reabsorption. At plasma concentrations of 1·4 mg/100 ml, 97–99·5% of the filtered ascorbic acid is reabsorbed. It appears in the urine as reduced (dehydro-) ascorbic acid, 2,3-ketogulonic acid, and as oxalate. Daily ingestion of 4 g or more of vitamin C increases the urinary excretion of oxalate, whereas daily intake of less than 4 g does not result in any increase.[16] About 40% of the urinary oxalate is derived from ascorbic acid but the mechanism of its formation is unknown[17] (Fig. 2).

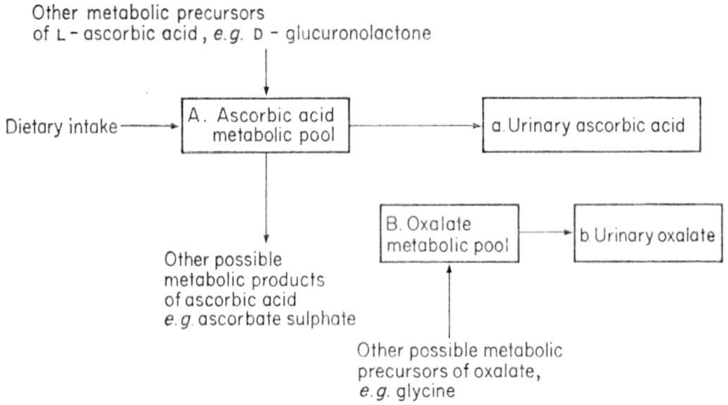

FIG. 2. Representation of the metabolism of ascorbic acid in man. The ascorbic acid metabolic pool in the plasma is derived from dietary intake of vitamin C and metabolic precursors during times of stress in guinea-pigs and humans. It may be transformed into metabolically-active products such as ascorbate sulphate. The plasma ascorbic acid appears in the urine as free ascorbic acid and also passes into the oxalate metabolic pool. 40% of the urinary oxalate is derived from ascorbic acid.

TISSUE SATURATION WITH VITAMIN C

To maintain tissue saturation in man the UK Panel on Recommended Allowances of Nutrients[8] stated that an intake of at least 60 mg/day is required. The panel indicated that there is no evidence indicating that man derives any benefit from such a high intake. Critics have claimed that the alleged large doses of the vitamin necessary for maintenance of tissue saturation may cause adverse effects. However, its presence is essential in adequate quantities to maintain normal metabolic function in the tissues. Physiological requirements differ greatly between individuals.[15,18] When metabolism is stimulated by a supply of the vitamin adequate to maintain tissue saturation,[19,20] the effect is demonstrated by greater gain in weight in more saturated animals, and by enhanced cell growth and anabolism (Fig. 3).

THE EFFECTS PRODUCED BY OVER-SATURATION WITH VITAMIN C

It has been pointed out by several investigators that many people are taking substantial amounts of supplementary vitamin C, and it is suggested that the doses which are being taken may cause adverse

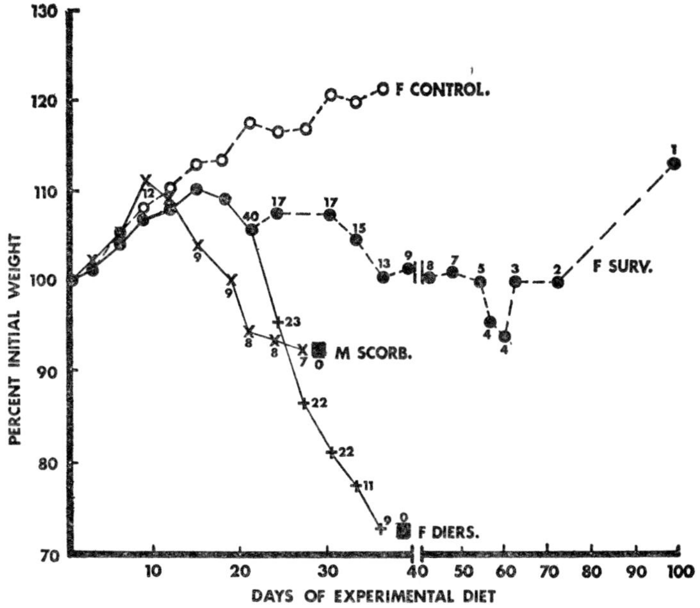

FIG. 3. The effect of vitamin C supplementation and deficiency on the growth and survival of guinea-pigs. The guinea-pigs received an ascorbic acid free diet from the first day. Female control guinea-pigs received 30 mg/kg vitamin C daily by intraperitoneal injection (F control). Male scorbutic guinea-pigs received the diet alone and all had died by day 28. On day 24 female scorbutic guinea-pigs were divided into those which were showing a rapid weight loss (F diers), and those which were maintaining their weight (F surv). F diers had all died by day 36 whereas F surv continued to survive for over 100 days. The numbers of survivors at each point in time are indicated (Odumosu and Wilson[85]).

side effects.[16,21,22] There is, however, very little definite evidence that any of these possible side effects actually occur, and the doses of vitamin C which have the capacity to produce them have not been specified or demonstrated to have these effects in humans. The side effects stated to result from administration of so-called large doses of supplementary vitamin C are shown in Table 1. It is noteworthy that the statements about the alleged side effects may greatly exaggerate the magnitude of the doses which have been recommended for prophylactic or therapeutic purposes.[34,38]

Gastro-intestinal disturbances have been reported in adults.[26] These are slight and may occur due to a laxative action of the vitamin C. The symptoms are less severe, and may be completely eliminated if the ascorbic acid is administered as sodium ascorbate. The

TABLE 1

Side effects stated to result from over-dosage with supplementary vitamin C in human beings

Reported side effects	*Stated daily dose vit. C*	*Reference*	*Comment*
1. Gastro-intestinal disturbances		23	Attributed to laxative effect[23] of vit. C
Increased peristalsis, abdominal colic	200–600 mg	24	
Gastro-enteritis and anal irritation	50–80 g	25	Reduced by oral administration[23] of Na ascorbate
Looseness of bowels	3 g	26	
Occasional diarrhoea	1 g	15	
2. Uric acid metabolism and gout			
Abnormal uric acid metabolism	1–2 g		No clinical evidence quoted in humans[22]
	Cold prophylaxis,	22	
Production of gout	Massive doses		
3. Oxaluria and stone formation			
Evaluation of urinary oxalate	4 g	27	Human metabolic evidence confirmed[17]
Formation of oxalate stones			No clinical evidence of stone formation[28]
4. Bone demineralisation			
Increased collagen catabolism	1 g (swine)	29, 30	No confirmatory evidence in humans[31]
Calcium resorption	5 mg/100 ml (chicken bone medium)		
5. Allergic symptoms			
Occasional urticaria and erythema		23, 32	Skin sensitivity confirmation
Transient urticaria	500 mg		Oral vit. C not confirmed as causative agent
Haemolytic crisis	High doses	33	Clinical confirmation of vit. C as causative agent
6. Human infertility	15–300 g	34	Unsupported by human tissue concentrations[36,37]
(Based on biochemical theory)	2–5 g	35	No definitive clinical evidence

frequency of this side effect is not greater than that of the signs of intolerance caused by other essential nutriments. Patients in Dublin have reported the occurrence of diarrhoea on rare occasions when taking doses of 0·5–1·0 g daily of supplementary vitamin C while suffering from cold symptoms. Such patients generally rapidly get over these symptoms.

The claim by Diehl[22] that certain groups of the population react to large doses of vitamin C by exacerbation of symptoms of gout appears to be unsubstantiated. No evidence suggests that ascorbic acid interacts with uric acid, and the effect of administration of supplementary vitamin C to patients with gout does not appear to have been studied. Oxalic acid is a metabolite of ascorbic acid but the urinary excretion of oxalic acid has been shown to be increased only when the daily intake of vitamin C exceeds 4 g. Briggs *et al.*[27] have confirmed that this dose, when administered for several days to one healthy individual, increased the urinary oxalate excretion ten-fold. Degradation of ascorbic acid through the oxalate pathway will be increased and the depressed urinary pH could lead to formation of oxalate stones. This might take place in individuals with abnormal oxalic acid metabolism although there is no evidence that it actually occurs.[28]

Ascorbic acid has been shown to play an essential role in the formation of both collagen and mucopolysaccharides which are principal components of the bone matrix. They appear to control bone mineralisation to some extent.[31] As a result of investigations in tissue cultures of chick bone, and in chickens[30] it has been claimed that large doses of supplementary vitamin C may have deleterious effects on bone metabolism in growing children. The concentration of ascorbic acid used in the chicken bone tissue culture medium was 5 mg/100 ml. The swine received 1000 mg supplementary vitamin C daily for 32 days. A level of 5 mg/100 ml ascorbic acid is about twice as high as any tissue levels of ascorbic acid reported in human beings even when they receive 10 g daily. It is difficult to compare the ascorbic acid levels in these species *in vivo* with those in humans. However, the supplementary vitamin C would elevate the already saturated tissue levels of these normal animals to excessive, and unnaturally, high values.

When excess ascorbic acid is available to the active cells in the bone matrix, relative protein deficiency, and related deficiencies in other components necessary for bone formation including iron, could lead to abnormal collagen formation and alterations in hydroxy proline excretion. Unless carefully controlled experiments are performed in order to eliminate the possibility of these relative deficiencies, it is unjustifiable to attribute any abnormalities which might appear solely to adverse effects of excess ascorbic acid rather than to induced deficiency of other essential factors. Brown[31] suggested that tissue concentrations of ascorbic acid existed in the chicken and swine tissues in the experiments which he described, comparable with those found in growing children to whom supplementary vitamin C had been administered during colds. However, he stated that there is no clinical evidence which confirms that administration of supplementary vitamin C produces bone demineralisation in children. During colds local and general desaturation of tissue ascorbic acid occurs. The object of prophylactic and therapeutic administration of vitamin C to children is to maintain normal tissue saturation with ascorbic acid, so that the maximal beneficial effect of the vitamin is obtained. In consequence, over-saturation with the vitamin is not produced with controlled therapy. Production of bone demineralisation by administration of excess supplementary vitamin C to children would be indicated by bone pains.[7] Bone pains have not been reported as a side effect in any of the trials in which supplementary vitamin C has been administered.[32–45]

Korner and Weber[23] stated that erythematous and urticarial reactions were observed only in exceptional cases following administration of supplementary vitamin C. Positive skin tests to ascorbic acid were reported in eight cases indicating that true allergic sensitivity to the vitamin can occur. A patient has also been reported who developed a haemolytic crisis whenever she took supplementary vitamin C in treatment of the common cold.[33] However, the frequency of these allergic effects, and the severity of the reactions, was not serious, and was no greater than the signs of intolerance caused by other essential nutrients.[23] Wilson and his colleagues (1974, unpublished observations) have received reports from school children of the occurrence of urticaria which lasted for 2–3 weeks at the beginning of term during oral administration of 500 mg of vitamin C daily. They did not have definitive evidence confirming that vitamin C was the causative agent of the urticaria. The urticaria was not severe and invariably cleared up completely.

It has been suggested that regular daily doses of vitamin C at 500–1000 times the usually accepted adult requirement, that is doses of 15–30 g daily (30 mg daily being the recommended British adult dose) may reduce fertility in women.[34] Subsequently Dr Briggs proposed that regular daily doses of 2–5 g vitamin C for a period of 14 months could lead to infertility.[35] Hoffer[36] has provided clinical evidence which demonstrates that supplementary doses of vitamin C in the range 3–30 g daily do not inhibit conception, and Briggs[37] has subsequently stated that there is no definitive clinical evidence that large doses of vitamin C can produce infertility.

PATHOLOGICAL STATES ASSOCIATED WITH TISSUE DEFICIENCY OF ASCORBIC ACID

Relative tissue desaturation of ascorbic acid may take place in metabolically active tissues if a sudden local tissue demand for ascorbic acid occurs during a period of stress. Reduction in efficiency and integrity of the desaturated tissues may then become apparent in the form of isolated scorbutic signs and symptoms, generally described as subclinical scurvy.[7] Some of the diseases in which local tissue demands are made for ascorbic acid are shown in Table 2. In all these disease states, signs and symptoms attributable to reduced tissue ascorbic acid concentrations can be found and a diagnosis of more or less severe subclinical scurvy can be made.

In diseases of the alimentary canal, the active process of absorption becomes inhibited when the mucosal cells are damaged either by disease or as a result of surgical interference with the bowel. The ascorbic acid concentrations of the plasma and leucocytes are reduced. Low ascorbic acid levels are found in the blood of patients with peptic ulceration and gastritis.[66] A defect in the absorption mechanism has also been reported in patients with ideopathic steatorrhoea.[50] It is probable that another factor, in addition to that involved in active or passive absorption of the vitamin, may be deranged.[7]

It has been shown by Hume *et al.*[46] that ascorbic acid concentrations are significantly reduced in the leucocyte after an acute myocardial infarction. At post-mortem, infarcted hearts had a higher tissue content of ascorbic acid than non-infarcted hearts. The evidence suggested that white blood cells emigrated to infarcted muscle carrying their complement of ascorbic acid where it was used in the inflammatory reaction and in order to promote healing. Wilson

TABLE 2

Disease states in which blood ascorbic acid levels are reduced and local tissue demands for ascorbic acid are increased

	Disease states	*References*
1.	*Peptic and duodenal ulceration*	7
	Intestinal malabsorption	46a
	Massive small bowel resection	47
	Sprue	48
	Ulcerated colitis	
	Ideopathic steatorrhoea	46a, 49, 50
	Malnutrition	51
2.	*Cardiovascular disorders*	
	Myocardial infarction	26
	Atherosclerosis	52, 53, 54
3.	*Respiratory disease*	
	Viral colds	26, 39, 55
	Allergic colds	42
	Bronchitis	
	Asthma	56
	Lung infections	51
4.	*Haemopoietic system*	
	Anaemia—iron deficient	6
	—pteroyl glutamic acid	49
	—pteroyl glutamic acid	57
5.	*Bone and joint disease*	
	Rheumatic fever	58, 59
	Rheumatoid arthritis	60, 61
	Slipped disc	62
	Osteoporosis due to iron overload	63
6.	*Cell growth*	
	Skin repair in burns	2
	Pressure sores	64, 65
7.	*Malignant cancerous states*	
	Gastric cancer	66
	Pulmonary carcinoma	67
	Acute lymphatic leukemia	67
8.	*Foetal development*	
	Teratogenesis	68
	Reduction in fertility and foetal growth	70

and Kevany[71] have demonstrated that there is an inverse relationship between blood cholesterol levels and tissue ascorbic acid concentrations in age groups above 40 years of age, and it has been suggested by Spittle[52] that atherosclerosis results from a long-term deficiency of vitamin C. In support of this hypothesis, Willis and Fishman[54] produced evidence from autopsy subjects that a localised depletion of ascorbic acid often exists in segments of arteries susceptible to atherosclerosis on account of mechanical stress. They suggest that it is possible to replenish ascorbic acid in arteries by administration of vitamin C, and Hume *et al.*[46] also indicate that vitamin C supplementation might be considered in patients with coronary thrombosis.

Plasma and leucocyte concentrations are significantly reduced during the common cold, and ascorbic acid utilisation is increased while symptoms are present.[26,30] There is also evidence that ascorbic acid concentrations are reduced in chronic bronchitis and that ascorbic acid metabolism[7,56] is altered in allergic disease. There is increased utilisation of ascorbic acid in order to maintain immunological defence mechanisms, and promote repair and growth of mucous membranes during and following the respiratory symptoms. For this reason, doses of supplementary vitamin C adequate to maintain normal leucocyte concentrations have been recommended for control of the respiratory symptoms.[39–45]

Patients with rheumatoid arthritis commonly have symptoms suggestive of infection and subclinical scurvy in addition to the joint pain and swelling which occurs in scorbutic patients.[7,72] Rinehart[58] has drawn attention to the possibility that vitamin C deficiency may play a role in the pathogenesis of rheumatic fever. Reduced plasma ascorbic acid levels occur in rheumatoid arthritis.[59–61] Ascorbic acid metabolism is abnormal in rheumatoid patients,[59] and it has been suggested that they should receive supplements of 200 mg daily,[73] particularly when they are also receiving aspirin.[61,74] Greenwood[62] has studied the effect of administration of 500–1000 mg vitamin C daily during several months for the preservation of the integrity of the spinal discs. He claims that this dose produces a therapeutic effect. The administration of an adequate dose to each patient is necessary in order to achieve benefit, and is influenced by stress, muscular activity, infection or injury.

Anaemia occurs in rheumatoid arthritis, in pregnancy,[69] in old age (Fig. 4) and in other conditions[75] in which ascorbic acid deficiency may occur. Ascorbic acid metabolism is closely linked with that of

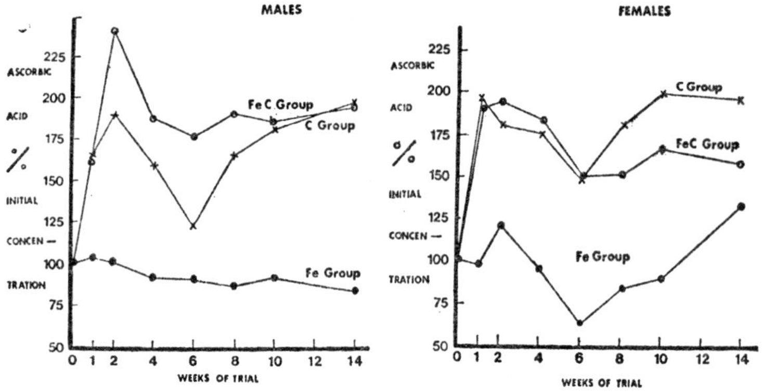

FIG. 4. Weekly changes in leucocyte ascorbic acid concentrations in male and female geriatric subjects following daily supplementation with 105 mg elemental iron (Fe groups), with vitamin C 500 mg (C groups), or with elemental iron and vitamin C together (Fe C groups) during 14 weeks. Note that leucocyte ascorbic acid concentrations consistently fall in the male Fe group whereas they rise in the female Fe group after week 6. They fall to a lower level in the male than in the female C group. Haemoglobin levels increased in all groups during the trial. No alteration took place in dietary intake of vitamin C in any of the groups (Loh and Wilson[6]).

iron.[4,75] 500 mg vitamin C should be administered with 100 mg active iron in order to prevent ascorbic acid deficiency in the tissues during iron therapy. Administration of B_{12}, folic acid[75] and pteroyl acid[57] makes excessive demands on ascorbic acid, and may in consequence lead to tissue desaturation. This gives rise to different types of anaemia the nature of which is determined by concurrent deficiency of other haemopoietic factors.

Ascorbic acid level has been shown to be elevated in recently burned human skin, and to be increased to a smaller degree in the edge of the burned area.[2] Ascorbic acid is concentrated in the inflamed area where leucocyte response is maximal and the initial metabolic changes are occurring in preparation for cell replacement and growth. Plasma ascorbic acid is reduced in gastric malignancy,[66] in lymphatic leukaemia,[77] and in secondaries of pulmonary carcinoma[67] (Table 3). It has been shown that ascorbic acid is concentrated in tumour tissue in animals.[78] It appears that the low plasma ascorbic acid concentrations occur because the ascorbic acid is being taken up by the tumour tissue and metabolised in the tumour cells.

Wilson and Loh[69] have employed supplementary vitamin C in doses of between 500 and 2000 mg daily to promote conception.

TABLE 3

Plasma, leucocyte, skin and secondary cancer ascorbic acid values in children and geriatric patients

	No. Patients	Leucocyte $\mu g/10^8$ cells	Plasma mg/100 ml
Children			
Control Blood	10	56·4 ± 21·9	0·40 ± 0·20
Leukemic blood	10	35·9 ± 15·1	0·95 ± 0·25
P		0·05	0·001
Geriatric			
Control blood	7	26·0 ± 17·1	0·47 ± 0·20
Lung Ca blood	1	12·5	0·13
		Normal mg/100 g	Tumour mg/100 g
Skin Biopsy	1	2·5	4·6

They have also provided evidence that ascorbic acid in the tissues is greatly reduced at the time of conception. They conclude that from the time of release of the ovum during fertilisation, during the first three months of pregnancy, and during the later stages of pregnancy, excessive demands are made on the mother and developing foetus for supplementary vitamin C. It has been demonstrated that the ascorbic acid concentration of venous umbilical blood is significantly higher than that of arterial umbilical blood or maternal venous blood at parturition[70] and that this ascorbic acid is made available for foetal metabolic requirements.[68a] If this ascorbic acid is not available for the mother she may develop signs of subclinical scurvy. According to Nelson and Forfar,[68] mothers who took supplementary vitamin C during pregnancy had half as many birth deformities in their newborn babies as were found in the control group. Deficiency of folic acid, or of ascorbic acid, during pregnancy may produce teratogenic effects. Wilson and Loh therefore propose[40] that pregnant mothers should receive a daily supplement of at least 500 mg of vitamin C together with the recommended daily dose of 100 mg of active iron.

TISSUE COMPENSATION TO OVER-SATURATION AND DESATURATION WITH VITAMIN C

The metabolic reaction to tissue over-saturation or desaturation with ascorbic acid is disputed in relation to the appearance of adverse side effects, and the development of compensatory tissue mechanisms. The experimental and clinical evidence indicates that tissue over-saturation with ascorbic acid is a rare phenomenon in normal humans on account of the limiting mechanisms in alimentary absorption and the efficiency of the renal excretory mechanisms for vitamin C. In pathological states, there is increased utilisation of ascorbic acid and it has proved difficult to give sufficient supplementary vitamin C by mouth to ensure the maintenance of normal leucocyte and plasma levels for periods longer than a few hours at a time. The suggestion that administration of supplementary vitamin C to children with colds, to patients with gastro-duodenal disease or steatorrhoea, with rheumatoid arthritis, with anaemia or with cancer (Refs. 32, 40, 42, 43, 45, 49, 50, 59, 66, 67, 73) can produce tissue over-saturation with consequent adverse effects attributable to the ascorbic acid has not been validated. Metabolic actions, and experimental effects attributed to ascorbic acid over-saturation have been reported after administration of doses of vitamin C which produce tissue levels considerably in excess of those ever found in diseased human beings. Administration of supplementary vitamin C to normal human beings does not appear to produce over-saturation effects in the laboratory.

Males and females do not differ in their maximum leucocyte ascorbic acid concentrations after receiving adequate supplementation.[79] Females have a lower intake of vitamin C than males.[80] In both sexes increased intake results in elevation of leucocyte concentrations but the sex differential between their blood levels is maintained.[81,82] Desaturation of tissue ascorbic acid associated with patho-physiological states is more pronounced in males than females.[83] It therefore appears that the sex difference in response to limited vitamin C supplementation, or to desaturation of ascorbic acid in response to patho-physiological states, is a metabolic characteristic of the cells *in vivo*. It has been shown that normal guinea-pigs are unable to synthesise vitamin C because they do not possess the enzyme gulonolactone oxidase which can convert gulonolactone to ascorbic acid. It has been assumed that human beings suffer from the same deficiency. Stone pointed out[84] that conclusive proof that both these species lacked the capacity to synthesise ascorbic acid was lacking.

Experimental evidence indicates that female guinea-pigs can survive for prolonged periods on scorbutogenic diets[85] (Fig. 3).

It has also been shown that gulonolactone prolongs life in guinea-pigs during vitamin C deficiency,[86] and that there is histochemical evidence of the presence of gulonolactone oxidase in the livers of scorbutic female guinea-pigs.[85] It therefore appears that female guinea-pigs which do not lose weight after day 24 of the scorbutogenic diet have the capacity to survive during the critical period until day 30 while they are conserving tissue ascorbic acid and activating paths of enzymatic synthesis. Increased levels of ascorbic acid then appear in the liver, and plasma ascorbic acid concentrations begin to rise. D-Glucuronolactone is converted to ascorbic acid in humans.[87,88] This suggests that humans also may be capable of synthesising the vitamin from suitable precursors.[17] Women develop less severe and fewer scorbutic symptoms than men in times of stress[5] and during vitamin C deficiency.[7] During tissue desaturation of ascorbic acid therefore, women, like female guinea-pigs, and possibly men, have the capacity to react to this form of stress by more economic utilisation, as shown by their reduction in metabolic utilisation, and by a limited degree of synthesis, of vitamin C.

REFERENCES

1. Ralli, E. P. and Sherry, S. (1941). *Medicine (Baltimore)*, **20**, p. 251.
2. Barton, G. M. G., Laing, J. E. and Barisoni, D. (1972). *Intern. J. Vit. Nutr. Res.*, **42**, p. 524.
3. Loh, H. S. and Wilson, C. V. M. (1973). *Intern. J. Vit. Nutr. Res.*, **43**, p. 355.
4. Loh, H. S. and Wilson, C. W. M. (1970). *Brit. J. Pharmacol.*, **40**, p. 169.
5. Wilson, C. W. M. (1973). *Vitamins, Editions Roche*, Hoffman La-Roche, Basle, **3**, p. 75.
6. Loh, H. S. and Wilson, C. W. M. (1971). *Lancet*, **1**, p. 110.
7. Wilson, C. W. M. (1974). *Practitioner*, in press, April.
8. HMSO, London (1969). *Reports on Public Health and Medical Subjects*, No. 120.
9. Odumosu, A. and Wilson, C. W. M. (1971). *Proc. Nutr. Soc.*, **30**, p. 81A.
10. Nicholson, J. T. L. and Chornock, F. W. (1942). *J. Clin. Invest.*, **21**, p. 505.
11. Loh, H. S. and Wilson, C. W. M. (1972). 5th *International Congress on Pharmacology*, Volunteer paper 847, p. 142.
12. Hornig, D., Weber, F. and Wiss, O. (1973). *Biochemical and Biophysical Research Communications*, **52**, p. 168.
13. Loh, H. S. and Wilson, C. W. M. (1970). *Brit. J. Pharmacol.*, **40**, p. 169.

14. Loh, H. S. and Wilson, C. W. M. (1971). *Intern. J. Vit. Nutr. Res.*, **41,** p. 445.
15. Wilson, C. W. M. and Greene, M. (1974). *Brit. J. Nutrit.*, (in press).
16. Lamden, M. P. and Chrystowski, G. A. (1954). *Proc. Soc. Exp. Biol. Med.*, **85,** p. 190.
17. Atkins, G. L., Dean, B. M., Griffin, W. J. and Watts, R. W. E. (1964). *J. Biol. Chem.*, **239,** p. 2975.
18. Booth, J. B. and Todd, G. B. (1970). *Brit. J. Hosp. Med.*, **4,** p. 513.
19. Evans, J. R. and Hughes, R. E. (1963). *Brit. J. Nutr.*, **17,** p. 251.
20. Williams, R. S. and Hughes, R. E. (1972). *Br. J. Nutr.*, **2,** p. 167.
21. Editorial (1973). *Brit. Med. J.*, **3,** p. 311.
22. Diehl, H. S. (1971). *Amer. J. Public Health.*, **61,** p. 649.
23. Korner, W. F. and Weber, F. (1972). *Intern. J. Vit. Nutr. Res.*, **42,** p. 528.
24. Wilson, M. G. and Lubschez, R. (1946). *J. Clin. Invest.*, **25,** p. 428.
25. Greer, E. (1955). *Med. Times*, **83,** p. 1160.
26. Hume, R. and Weyers, E. (1973). *Scot. Med. J.*, **18,** 3–7.
27. Briggs, M. H., Garcia-Webb, P. and Davies, P. (1973). *Lancet*, **2,** p. 201.
28. Kean, W. F. (1974). *Lancet*, **1,** p. 364.
29. Brown, R. G., Sharma, V. D. and Young, L. G. (1971). *Can. J. Anim. Sci.*, **51,** p. 439.
30. Ramp, W. K. and Thornton, P. A. (1971). *Proc. Soc. Exp. Biol. Med.*, **132,** p. 618.
31. Brown, R. G. (1973). *J. Amer. Med. Ass.*, **224,** p. 1529.
32. Wilson, C. W. M., Loh, H. S. and Foster, F. G. (1973). *Europ. J. clin. Pharmacol.*, **6,** p. 196.
33. Goldstein, M. L. (1971). *J. Amer. Med. Ass.*, **216,** p. 332.
34. Briggs, M. H. (1973). *Lancet*, **2,** p. 677.
35. Briggs, M. H. (1973). *Lancet*, **2,** p. 1083.
36. Hoffer, A. (1973). *Lancet*, **2,** p. 1146.
37. Briggs, M. H. (1973). *Lancet*, **2,** p. 1711.
38. Wilson, C. W. M. and Loh, H. S. (1969). *Fourth International Congress on Pharmacology*, Basle, Abstracts of Communications, 458.
39. Wilson, C. W. M. and Loh, H. S. (1969). *Acta Allergolica, XXIV*, p. 367.
40. Wilson, C. W. M., Loh, H. S. and Foster, F. G. (1973). *Europ. J. clin. Pharmacol.*, **6,** p. 26.
41. Wilson, C. W. M. and Loh, H. S. (1974). *Europ. J. clin. Pharm.* (in press).
42. Wilson, C. W. M., Loh, H. S. and Greene, M. (1973). *Europ. Nutr. Conf. Cambridge*, Abstract No. 49 (in press *Brit. J. Nutr.*).
43. Anderson, T. W., Reid, D. B. W. and Beaton, G. H. (1972). *Canad. Med. Assoc. J.*, **107,** p. 503.
44. Cowan, D. W., Diehl, H. S. and Baker, A. B. (1942). *J. Amer. Med. Ass.*, **120,** p. 1267.
45. Coulehan, J. L., Reisinger, K. S., Rogers, K. D. and Bradley, D. W. (1974). *New Engl. J. Med.*, **290,** p. 6.

46. Hume, R., Weyers, E., Rowan, T., Reid, D. S. and Hills, W. S. (1972). *Brit. Heart Journal*, **34,** p. 238.
46a. Williamson, J. M., Goldberg, A. and Moore, F. M. L. (1967). *Brit. Med. J.*, **1,** p. 23.
47. Hughes, R. E. (1973). In *Molecular Structure and Function of Food Carbohydrate*, Eds. G. G. Birch and L. F. Green, Applied Science, London, p. 108.
48. Farmer, C. J., Abt, A. F. and Chirm, H. (1940). *Bull. Northwestern Univ. Med. Sch.*, **14,** p. 114.
49. Boscott, R. J. and Cooke, R. T. (1954). *Quart. J. Med.*, **23,** p. 307.
50. Stewart, J. S. and Booth, C. C. (1964). *Clin. Sci.*, **27,** p. 15.
51. Gupta, S. and Santhanagopalan, T. (1964). *Indian Pediatrics*, **1,** p. 361.
52. Spittle, C. R. (1971). *Lancet*, **2,** p. 1280.
53. Ginter, E. (1972). *Lancet*, **1,** p. 1233.
54. Willis, G. C. and Fishman, S. (1955). *Canad. Med. Ass. J.*, **72,** p. 500.
55. Loh, H. S. and Wilson, C. W. M. (1971). *Nutr. Reports Intern.*, **41,** p. 371.
56. MacNally, N. (1953). *J. Irish Med. Assoc.*, **33,** p. 175.
57. Laheri, S. and Banerjee, S. (1956). *Proc. Soc. Exp. Biol. Med.*, **91,** p. 545.
58. Rinehart, J. F. (1935). *Ann. Intern. Med.*, **9,** p. 586.
59. Rinehart, J. F., Greenberg, L. D. and Baker, F. (1936). *Proc. Soc. Exp. Biol. Med.*, **35,** p. 347.
60. Abrams, E. and Sandson, T. (1964). *Am. Rheum. Dis.*, **23,** p. 295.
61. Sahud, M. A. and Cohen, R. J. (1971). *Lancet*, **1,** p. 937.
62. Greenwood, J. (1964). *Medical Annals of the District of Columbia*, **33,** p. 274.
63. Lynch, S. R., Berelowitz, I., Seftel, H. C., Miller, G. B., Krawltz, P., Charlton, R. W. and Bothwell, T. H. (1967). *Amer. J. Clin. Nutr.*, **20,** p. 799.
64. Burr, R. G. and Rajan, K. T. (1972). *Brit. J. Nutr.*, Sept., p. 275.
65. Lloyd, J. V., Davis, P. S., Emery, H. and Lander, H. (1972). *J. Clin. Path.*, **25,** p. 478.
66. Freeman, J. T. and Hafkesbring, R. (1957). *Gastroenterology*, **32,** p. 878.
67. Kakar, S. C. and Wilson, C. W. M. (1974). *Proc. Brit. Nutr. Soc.*, (in press).
68. Nelson, M. M. and Forfar, J. O. (1971). *Brit. Med. J.*, **1,** p. 523.
68a. Teel, H. M., Burke, B. S. and Draper, R. (1938). *Am. J. Disease Children*, **56,** p. 1004.
69. Wilson, C. W. M. and Loh, H. S. (1973). *Lancet*, **2,** p. 859.
70. Fukuda, T. and Horiuchi, S. (1959). *J. Vitaminol.*, **5,** p. 151.
71. Wilson, C. W. M. and Kevany, J. P. (1972). *Brit. J. prev. soc. Med.*, **26,** p. 53.
72. Buchanan, W. W. and Dick, W. C. (1972). *Medical Laboratory Technology*, **29,** p. 109.
73. Hall, M. G., Darling, R. C. and Taylor, F. H. L. (1939). *Ann. Intern. Med.*, **13,** p. 415.

74. Loh, H. S. and Wilson, C. W. M. (1973). *J. Clin. Pharmacol.*, **13,** p. 480.
75. Cox, E. V. (1968). *Vitamins and Hormones. Advances in research and applications,* Ed. R. S. Harris, Vol. **26,** Academic Press, London, p. 635.
76. Odumosu, A. and Wilson, C. W. M. (1971). *Brit. J. Pharmacol.*
77. Kakar, S. C., Wilson, C. W. M. and Bell, J. M. (1974). *Unpublished observations.*
78. Voegthin, C., Kahler, H. and Johnson, J. M. (1937). *Am. J. Cancer*, **29,** p. 477.
79. Loh, H. S. and Wilson, C. W. M. (1971). *Brit. Med. J.*, **3,** p. 733.
80. Morgan, A. F., Gillum, H. L. and Williams, R. I. (1955). *J. Nutrition.*, **55,** p. 431.
81. Woodhill, J. M. (1970). *Intern. J. Vit. Res.*, **40,** p. 520.
82. Loh, H. S. (1972). *Intern. J. Vit. Nutr. Res.*, **42,** p. 86.
83. Loh, H. S., Odumosu, A. and Wilson, C. W. M. (1974). *Clin. Pharm. Therap.* (in press).
84. Stone, I. (1966). *Perspect. Biol. Med.*, **10,** p. 133.
85. Odumosu and Wilson (1973). *Nature.*
86. Odumosu and Wilson (1973). *Proc. Nutr. Soc.*
87. Goldstone, A. and Adams, E. J. (1962). *J. Biol. Chem.*, **237,** p. 3476.
88. Baker, E. M., Sauberlich, H. E., Wolfshill, S. J., Wallace, W. T. and Dean, E. E. (1962). *Proc. Soc. Exp. Biol. Med.*, **109,** p. 307.

15

Recent Advances in the Molecular Biology of Vitamin C†

S. Lewin

Department of Postgraduate Molecular Biology,
North East London Polytechnic,
Romford Road, London, England

ABSTRACT

The considerable evidence in favour of the beneficial effects of high intake of ascorbic acid in many individuals can be explained in terms of (*a*) *direct involvement of the ascorbate system, and* (*b*) *an indirect impact by virtue of its enhancing the activities of* 3′,5′ *cyclic AMP and* 3′,5′ *cyclic GMP. The direct activity can be understood in terms of the physico-chemical properties of the ascorbic acid and dehydro-ascorbic entity comprising oxidation–reduction activity, complex formation, H . . . bonding, and lowering of the interfacial tension. The indirect activity is considered to proceed via various hormone-receptor systems with cyclic AMP and cyclic GMP acting as 'second messengers'. In the case of cyclic AMP, ascorbate enhances formation of adrenaline protected from oxidative degradation; the adrenaline potentiates adenyl cyclase which in turn activates the formation of cyclic AMP from ATP. The ascorbate also inhibits phosphodiesterase hydrolytic breakdown of cyclic AMP. The inhibition of phosphodiesterase extends also to enhancing an increase in the activity of cyclic GMP by protecting the latter from hydrolytic degradation. Both cyclic AMP and cyclic GMP potentiate numerous physiological activities which encompass enzyme production, secretions, increased membrane permeability, interferon activation and other physiological activities. It is considered that balanced concentration ratios of* [(*cyclic AMP*)/(*cyclic GMP*)] *result in increased resistance to bodily malfunctions.*

Theoretical considerations and experimental evidence have been accumulated showing that the free ascorbate ion is not only more readily oxidised but is also inherently unstable, anaerobically, compared to the acid form, as the ring structure is hydrolytically ruptured and the double bond is hydrolytically replaced by a single bond. Both ring structure and open chain forms possess reducing properties and this parameter can result in a source of error in assays of the biologically active ring-structured vitamin C, when reduction capacities are involved.

† The information presented in this paper is based on part of the author's book entitled *The Molecular Biology and Potential of the Ascorbic System*, (Academic Press, in press, 1975).[1]

INTRODUCTION

Considerable evidence has been accumulated over the last three to four decades that intake of mega quantities of ascorbic acid—nutritionally termed vitamin C—results, in numerous individuals, in significant beneficial effects over a wide range of biochemical stresses and associated body malfunctions. This has led to the thesis that the smaller doses—in the range of 50–100 mg total daily dose—medically recommended for avoidance of the clinical symptoms of scurvy, are unlikely to meet the optimal requirements.

Several classes of arguments have been advanced in favour of this thesis. Some are concise; others were based rather on faith and on non-strict parallelisms. The arguments in favour of this thesis can be conveniently marshalled on the following lines.

First, the combination of experimental findings that individuals, or animals, who undergo biochemical stress and body malfunctions are characterised by depleted levels of vitamin C—in particular tissues, *e.g.* leucocytes—as compared with those not subject to the upsets (*e.g.* Refs. 2, 3, 4); for reviews *see*, for example, Refs. 5, 6, 7); and, further, that when this deficiency is rectified by high intake of ascorbic acid (as evidenced in increased tissue vitamin C levels) the increased levels of the vitamin in the blood (*e.g.* Refs. 8, 9) are paralleled by increased resistance to the malfunction (*e.g.* Refs. 10, 11).

Second, the presence of high concentrations of vitamin C in several tissues concerned with response to biochemical stress (such as the adrenals, leucocytes and pancreas)—for relevant values *see* Table 2 in Ref. 12—suggests both immediate use and storage as required in the potentiating of particular molecules which may be involved in hormonal paths.

Third, animals synthesising their own vitamin C have been shown to increase their synthesis about ten-fold when subjected to biochemical stress, such as drugs (*e.g.* Ref. 13). It follows that if animals which synthesise their own vitamin C switch over to a multifold production, on exposure to biochemical stress, it must be so because they need the extra quantity. Those animals—including man—who cannot meet their vitamin C requirements by synthesis should, on being subjected to biochemical stress, ingest a corresponding multifold amount of the vitamin; otherwise their physiology would be adversely affected.

Once we accept these considerations, several questions naturally arise: How can a small molecule such as vitamin C potentiate so many physiologically beneficial activities? What are the paths via which the vitamin can express its potential? In attempting to answer these and associated questions it is possible to utilise two different approaches which nevertheless utilise the same available material. One approach is to collect and classify all the experimental evidence concerning the effect of intake of mega quantities of vitamin C, and then attempt to construct a pattern which would be admissible stereochemically and energetically in respect of the potential of the ascorbic system. A second approach is to consider first the potential of all the relevant physico-chemical characteristics of the ascorbic system and then evaluate the impact they could make on the biological system by virtue of exerting specific influences on its various physiological activities.

Now, considerable documentation and classification of the various physiological influences of vitamin C have been undertaken in various fields. Some fields have been well explored and explained, but a common relationship was still not forthcoming. For example, the participation of the ascorbate system in a number of enzymatic reactions, such as in the hydroxylation of lysine and proline in the formation of collagen and in the hydroxylation of thymidine (for review *see* Ref. 14) and in the formation of adrenaline from dopamine (*e.g.* Ref. 15) have been evaluated well onto the molecular level. However, these activities cover only a small section of the wide range of ascorbate-enhanced physiological activities, which range over such widely different fields as coronary malfunction, the common cold, cancer, chemical toxicity and allergy. The common denominator(s) until recently appeared elusive.

As a result of extensive examination and re-examination of the overall field, and using the second approach to the solution of the *modus operandi* of the ascorbic system, I have come to the conclusion that the beneficial effects of intake of mega quantities of ascorbic acid can be conveniently evaluated if the resultant activities are considered in two distinct categories, namely activities which result from direct action, and activities which result from secondary actions via hormonally potentiated paths. The direct action potential can be evaluated using the physico-chemical characteristics of the ascorbic system, while the indirect action involves potentiation of cyclic AMP and cyclic GMP and consequent evaluation of relevant hormonal paths accompanied by associated physiological activities which extend the ability of the biological system to resist the encountered biochemical stresses on a very wide front.

(A) PHYSICO-CHEMICAL CHARACTERISTICS AND THEIR DIRECT POTENTIAL

(I) Structure and Derivatives

Overall view of the ascorbic system is facilitated by a brief outline of its structure and derivatives.

As a result of the investigations of the Haworth/Hirst school,[16–20] the accepted formula of L-ascorbic acid can be represented by

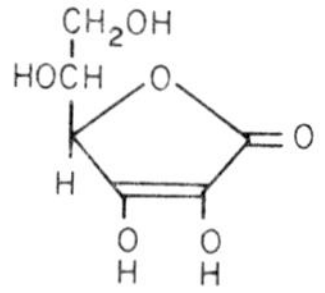

It is desirable to outline at this stage various possible interactions of ascorbic acid. These are given in Fig. 1.

(II) Physico-Chemical Characteristics

These are outlined as follows.

(i) *Acid–base ionisation*

Ascorbic acid can ionise in two stages involving the two p*K* values

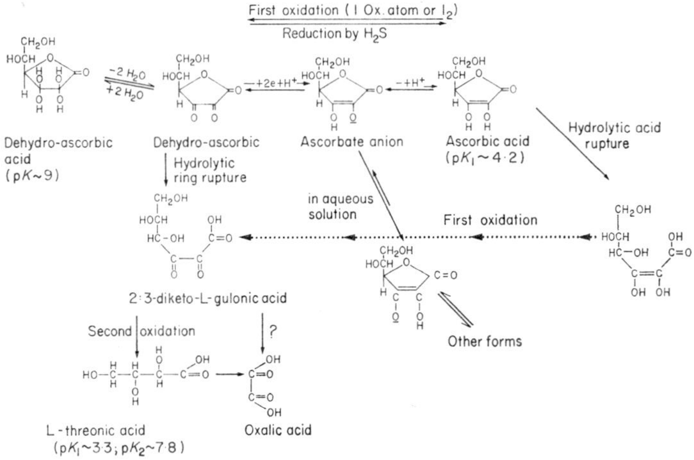

FIG. 1. Ascorbic acid and derivatives.

of 4·18 and 11·6 (at 37°C).[21,22] Hence, at the physiological pH range of 7·2–7·4 the ascorbic molecule is an anion with a single negative charge.

(ii) Cation-complexing potential

The ascorbate anion can complex with divalent cations, thus

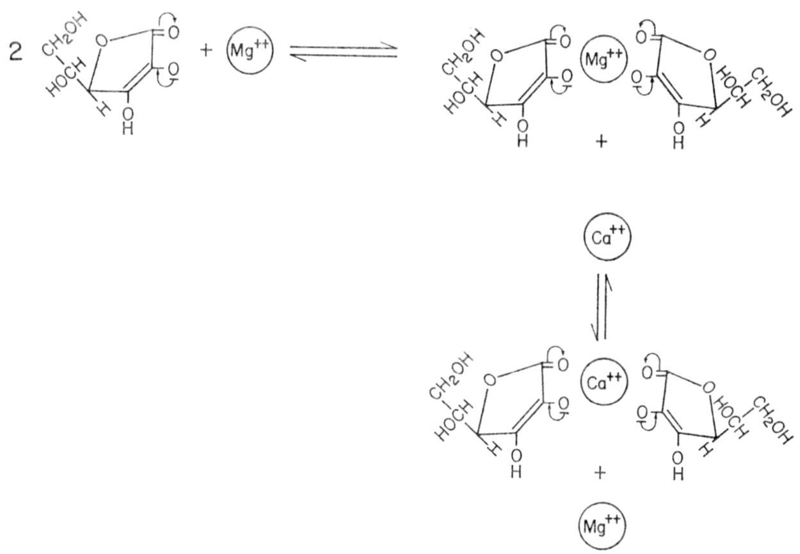

(iii) H . . . bonding

The $>$C—O^{-}, the $>$C=O and the $>$C—OH groups can act as H . . . acceptors, while the $>$C—OH group can act as an H . . . donor.

These characteristics and the resonance potential entailed in the double-bond between the α-C and the β-C endow this section of the ascorbic molecule with an enhanced potential for both unidirectional and complementary-two-directional double H . . . bonding activities, thus:

(*a*) *Unidirectional double-H . . . bonding*. This can be expressed in interaction between the guanidinium group of arginine and the ascorbate. This is diagrammatically illustrated in Fig. 2a, and using correctly proportioned space-filling molecular models, in Fig. 2b.

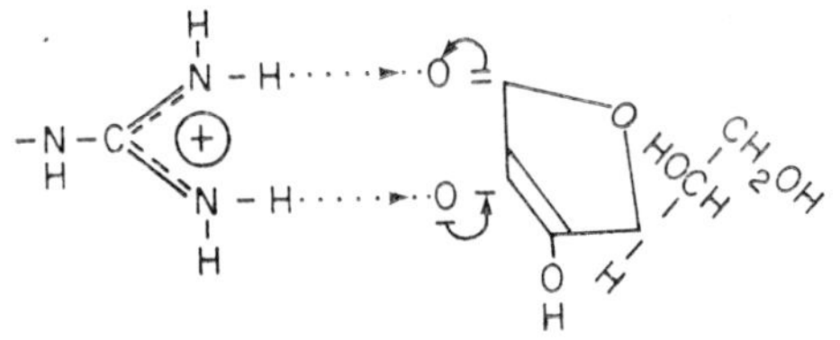

FIG. 2 (a).

(*b*) *Complementary double-H . . . bonding.* Using correctly-proportioned space-filling molecular models, it can be readily shown that the α-C—O⁻ and the adjoining β-C—OH group can participate in complementary H . . . bonding with corresponding H . . . donor and H . . . acceptor groups suitably located in another molecular species. This can be postulated to result in the formation of an association between the 3′OH and 4′OH groups of adrenaline or of noradrenaline and the relevant section of the ascorbate entity. Diagrammatically this can be represented as in Fig. 3a, and as a molecule-model in Fig. 3b.

(*iv*) *Redox potential associated with oxidation/reduction activity of the ascorbic system*

The ascorbic system is involved in oxidation/reduction activity, thus

$$\text{Ascorbic acid} \underset{}{\overset{-2e-2H^{+}}{\rightleftharpoons}} \text{Dehydro-ascorbic acid (DHA)}$$

The system is diagrammatically represented at the top of Fig. 1.

The redox potential of the system is known to have the value of +0·058 V at pH 7·0 and (25–30°C).[23] The pH value of blood of normal persons is in the region of 7·4 (and 37°C). The value of the redox potential under these conditions is somewhat lower, approximately by 10 mV.

(*v*) *Optical properties and pH values*

Neutral solutions of sodium ascorbate (pH range of *ca.* 7·2–7·7, when freshly prepared) display certain characteristics and changes with respect to optical absorbance, optical rotation and pH values. In concentrations of 10^{-2}M and higher the optical rotation values and the optical absorbance values remain steady for several days at 37°C.[1] However, the presence of oxygen, cations with more than one valency, rise in temperature and increasing dilution result in changes

FIG. 2 (b).

FIG. 2. Unidirectional H-bonding in guanidinium ascorbate. (a) Diagrammatic representation; (b) molecular model representation using correctly-proportioned space-filling atomic models (CPK).

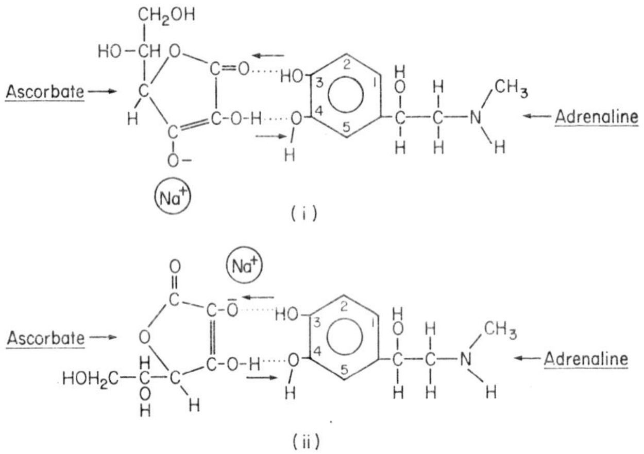

FIG. 3 (a).

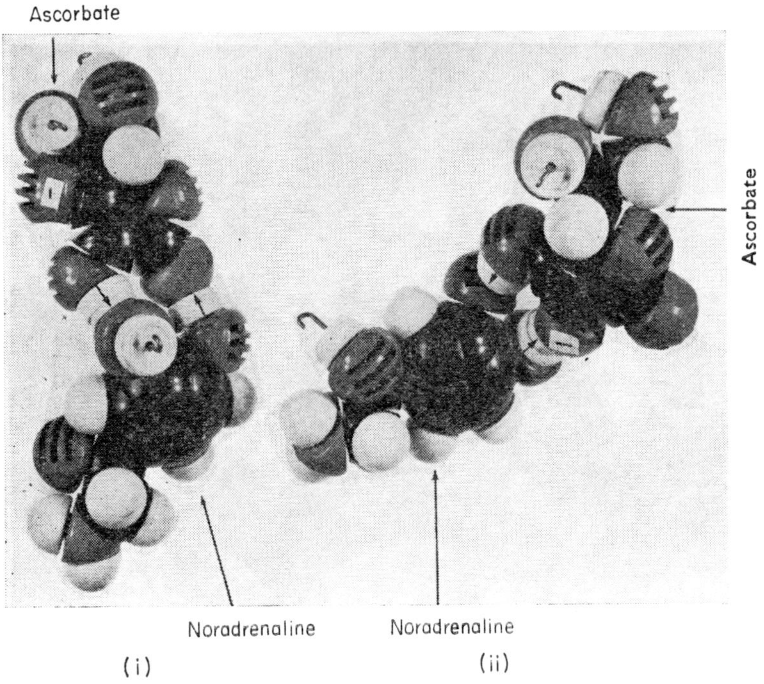

FIG. 3 (b).

FIG. 3. Complementary H-bonding between adrenaline and ascorbate. (a) Schematic representation; (b) molecular model representation using correctly-proportioned space-filling atomic models (CPK).

indicative of the destruction of the ascorbate. The effect of time on the stability of the ascorbate entity can be followed by the decrease in optical density and by the decreasing values of the optical rotation. Figure 4 illustrates the changes taking place in the optical density of dilute ascorbate solutions under anaerobic conditions. The optical changes in dilute ascorbate solutions can be avoided by an increase in ionic strength—using a neutral salt such as sodium chloride—to regions of 10^{-2}.[1]

In dilute aqueous ascorbate solutions, we have noted small pH changes[1] which can be attributed to a change in the site of acid ionisation from that of the α-carbon OH group to the OH group of the β-carbon, which change would reduce the electrostatic repulsion between the negatively charged O^- group and the negative end of the dipole of the $>C^{\delta+}=O^{\delta-}$ group.

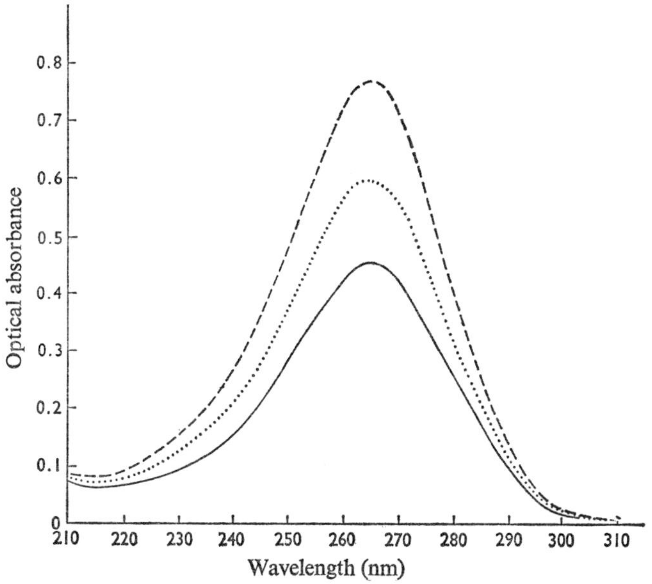

FIG. 4. Effect of time on the optical absorbance spectrum of 50 μm-sodium ascorbate, pH 6·1 25°C, by using 1 cm optical path length under anaerobic conditions. At 9 min after preparation of the solution (- - - -); 70 min after preparation of solution (· · · ·); 130 min after preparation of solution (——). The solution was prepared directly from equivalent quantities of pure ascorbic acid and $NaHCO_3$ while bubbling N_2, free from O_2 and CO_2, by using pure-conductivity water.

In this connection the saline content of serum (namely *ca.* 0·8 %) which is approximately 0·2M can be computed to protect the dilute ascorbate in the serum, *i.e.* in a concentration of *ca.* 10^{-4}M. It is relevant to point out here that it has been computed—using optical density values as the criteria of stability of the double-bonded (α-C to β-C) γ-lactone structure of the biologically active vitamin C—that it is more advantageous to ingest the vitamin in solutions of grapefruit juice or orange juice than in water, because of the protective influences of the juices.[1,24,25]

(*vi*) *Surface tension reducing activity*

It has been computed on theoretical grounds that ring-structures—by decreasing the entropy of the neighbouring/associated water molecules—should reduce the interfacial tension.[1,26] The ascorbic system by virtue of its γ-lactone ring structure should conform to this pattern; experimentally this is the case.[1]

(III) The Potential of Physico-Chemical Activities of the Ascorbic System

Let us now evaluate certain characteristics of the ascorbic system which can result in direct beneficial influence on mechanisms involved in resisting biochemical stress.

(*i*) *Vascular stress and toxic stress*

Biochemical stress on the vascular system can be attributed to a number of causes. Restricting ourselves to two factors, namely the formation of arterial deposits and the need to keep down the levels of Na^+ in the serum and the removal of toxic cations, we can apply the following considerations.

(*a*) *Arterial deposits.* Arterial 'furry' deposits can be considered as insoluble complexes of calcium/phospholipid/cholesterol. Such complexes can be dissolved readily by comparatively high concentrations of ascorbate. I have noted that aged artificial precipitates of such deposits (made by mixing individual solutions of calcium salts, phospholipids and cholesterol and allowing them to age) tend to be re-dissolved by shaking vigorously with large volumes of sodium ascorbate/physiological saline solutions (which contain *ca.* 20 mg ascorbate per 100 ml).

These results are in accord with the proposed suggestion that higher ascorbate concentrations in the serum would tend to redissolve the furry arterial deposits by two complementary activities, namely.

(1) Small lowering of the surface tension of the serum with consequent enhancement of the removal of the cholesterol constituent from the arterial deposits,† and

(2) removal of the calcium from the arterial deposits, thus

$$2\ \text{Ascorbate}^-\text{Na}^+ + \left\{\begin{matrix}\text{Ca}^{++},2(\text{phospholipid})\\ \text{cholesterol}\end{matrix}\right\}_{\textit{Insoluble}} \rightleftharpoons \underset{\textit{Soluble}}{2\ \text{Ascorbate}^-,\text{Ca}^{++}} + \underset{\textit{Soluble}}{2\ \text{Na}^+\text{phospholipid}^-}$$

$$\longrightarrow$$

(*b*) *Removal of* Na^+ *and of toxic cations.* Intake of mega quantities of ascorbic acid can be computed to enhance removal of Na^+ via the urine, thereby reducing the level of sodium ions in the serum, as is outlined in the following considerations.

The pH of urine in humans varies between 6·4 and 8·2 with an average cluster of values at about pH 6 (*e.g.* Refs. 29–32). At this pH value vitamin C is almost entirely in the singly charged anionic form while accompanied by a positive counterion; and this applies to the excreted material. Most of the cations excreted in the urine are metallic, with Na^+ having the highest concentration. It can be computed that when about 3 g of vitamin C are excreted in the urine, about 0·4 g of additional Na^+ should be co-excreted, thereby benefiting patients in which the level of Na^+ should be reduced (*e.g.* patients with coronary malfunction).

It should be noted that in addition to enhanced excretion of Na^+, enhanced excretion of K^+, NH_4^+, Ca^{++}, Mg^{++}, iron, copper and zinc are to be expected although to a much lower extent. The latter enhancement is being currently monitored in a number of

† The solubility of cholesterol rises with decrease in the surface tension of the solvent medium. The surface tension of serum/air at 37°C is *ca.* 47 dyn/cm.[27]

The presence of ascorbic acid/ascorbate lowers the surface tension of water/air by over 20 dyn/cm (*e.g.* Ref. 28) and that of serum/air by several dynes.[1]

volunteers. Further, the highly toxic Pb^{++}, Hg^{++}, Cd^{++} and radioactive Sr^{++} should also display enhanced excretion. Indeed, experimental evidence exists[33–35] that the intake of mega quantities of vitamin C results in amelioration and increased resistance to the toxic effects following ingestion of mercury and lead; this could be due, at least in part, to enhanced removal as cationic partners of the ascorbate anion in the urine.

(*ii*) *Increased resistance to bacterial and viral infections*

(*a*) *Increased resistance to bacterial infection.* Bacterial infections are resisted by antibodies most of which are chemically classifiable as γ-globulins. γ-Globulins are composed of proteins which comprise a significant number of disulphide bridges in linking their light and heavy chains. When the body is subjected to bacterial attacks, the anti-bacterial activity involves increased production of γ-globulins. The formation of γ-globulins is genetically restricted to initial formation of chains with sulphydryl groups which are only subsequently oxidised to the disulphide bridges. The genetic restriction arises because no tri-nucleotide is known to code for *cystine*; only *cysteine* is coded for by UGU and UGC.[12] Now, the ingested food contains proteins with a preponderance of constituent *cystines*. The required increased formation of cysteine can be met only when the existing cystines are correspondingly reduced to cysteines. This can be readily carried out by increasing the concentration of the powerful reductant ascorbate, following corresponding mega intake of vitamin C. The resultant DHA can be subsequently used to re-oxidise the SH groups of the newly biosynthesised globulin protein chains thereby making possible increased production of immunoglobulin. The increased oxidation/reduction activity following the intake of mega quantities of vitamin C can obviously apply to several other physiological activities which involve oxidation/reduction paths.

(*b*) *Increased resistance to viral infection.* Viruses consist each of nucleic acid surrounded by nucleoprotein, often associated with an external layer of lipid. The ascorbic system when present in concentrations such as 10^{-3}M—a value easily over-reached in leucocytes, adrenals and other tissue—can therefore enforce a lowering of several dynes of the interfacial tension. At a cellular/surrounding liquid interfacial tension of *ca.* 3–10 dyn/cm, this would constitute a major drop in the value of the hydrophobic adherence equilibrium constant.[26,36,37] Since hydrophobic adherence contributes substantially to protein–protein, protein–lipid and lipid–lipid associations on which the virus relies for protection of its nucleic acid from attack by body nucleases, such lowering of the hydrophobic adherence can readily result in dissolution of the protective protein/lipid sheaths, followed by adverse effects on the viral nucleic acid, thereby weakening its attack on the human.

(B) INDIRECT ACTIVITY USING CYCLIC AMP AND CYCLIC GMP AND THEIR POTENTIATION OF HORMONAL ACTIVITIES

(I) General

The above considerations regarding the beneficial effect of intake of mega quantities of vitamin C in respect of biological activities cover only a fraction of the field. Now human physiology is greatly activated/controlled by hormone action. In turn hormone action is potentiated in large measure by the cyclic nucleotides c-AMP and c-GMP[38,39] and their mutual antagonism, *e.g.* in cellular proliferation and differentiation.[40] This indicates that vitamin C may well be involved in paths which increase the potential/concentrations of the two cyclic nucleotides. This indication is backed by the following coincident activities which implicate the cyclic nucleotides and vitamin C as associates in potentiating various hormonal activities.

(*i*) *Cyclic AMP*

(*a*) *Cancer*. Ascorbate was found to be highly toxic to Ehrlich ascites-carcinoma cells.[41] Ascorbate concentrations are depleted in malignant activity (*see* for example Ref. 3); diminished adenyl cyclase activity has been noted in polyoma-virus-transformed cells.[42] Tumour growth was inhibited by cyclic AMP.[43]

Cell growth *in vitro* can be inhibited by cyclic AMP.[44]

(*b*) *RNA transcription*. Price[45] observed that ascorbate can replace thiol reducing agents in RNA transcription *in vitro*; cyclic AMP was shown to activate *lac* mRNA transcription in *E. coli* when cyclic AMP binding protein was present (*e.g.* Ref. 46).

(*c*) *ACTH*. In the adrenal cortex cyclic AMP has been shown to mimic the effects of ACTH on ascorbate depletion.[47]

(*d*) *Cyclic AMP and interferon production.* The presence of 1mM cyclic AMP results in significant antiviral activity in chick fibroblasts; this contrasts with the absence of antiviral activity by cyclic AMP alone, and the inability of 5′AMP to potentiate interferon activity.[48] The trend of high intake of vitamin C to combat the effects of the common cold may act indirectly via increased interferon activity, provided ascorbate enhances cyclic AMP action. It will be seen later that experimental evidence does exist that ascorbate addition *in vitro* can potentiate cyclic AMP activity.

(*ii*) *Cyclic GMP*

Experimental evidence has been accumulated showing that insulin potentiates cyclic GMP activity on adipose and liver cells.[49]

Intravenous intake of vitamin C in 0·5 to *ca.* 1·2 g has been shown to result in lowering of the blood sugar curves in normal and diabetic patients.[50,51] Further, intake of vitamin C was found to reduce the insulin dose required by diabetic patients. This relation was observed by Pfleger and Scholl,[52] Bartelheimer and others;[53,54] and Dice and Daniel[55] concluded that in one diabetic patient intake of each gramme of the vitamin resulted in two fewer units of insulin being required by the patient.

(II) Suggested Mechanism for the Effect of High Intake of Vitamin C on the Activities of Cyclic AMP and Cyclic GMP

The pattern of activity advocated here is that high intake of vitamin C enforces increasing levels of activity of both cyclic AMP and cyclic GMP.

In respect of cyclic AMP it is proposed that the influence of ascorbate is two-pronged, namely (1) enhanced protection in the formation of adrenaline which potentiates adenyl-cyclase (present in the cell-membrane) to activate the formation of cyclic AMP—from ATP—and (2) inhibition of phosphodiesterase (PDE) from hydrolysing the cyclic AMP present to 5′-AMP. In contrast, the activity of cyclic GMP is considered to be enhanced only as a result of inhibition by ascorbate of the phosphodiesterase hydrolysing cyclic GMP to 5′-GMP.

The various stages and overall involvements are shown schematically in Fig. 5a and Fig. 5b. Each vitamin C involvement will now be considered in turn.

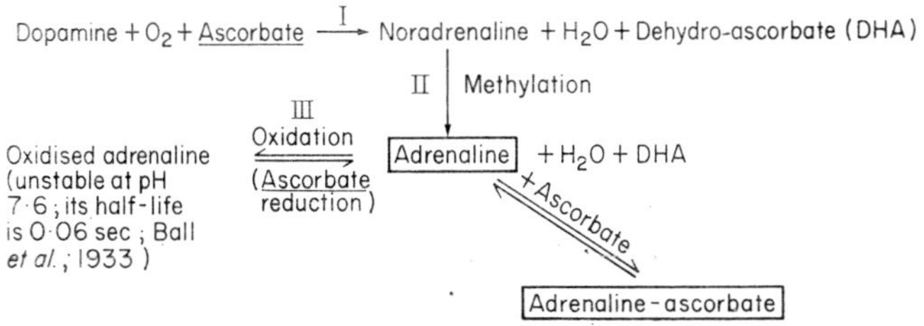

FIG. 5a. Schematic representation of the ascorbate-assisted formation of adrenaline from dopamine.

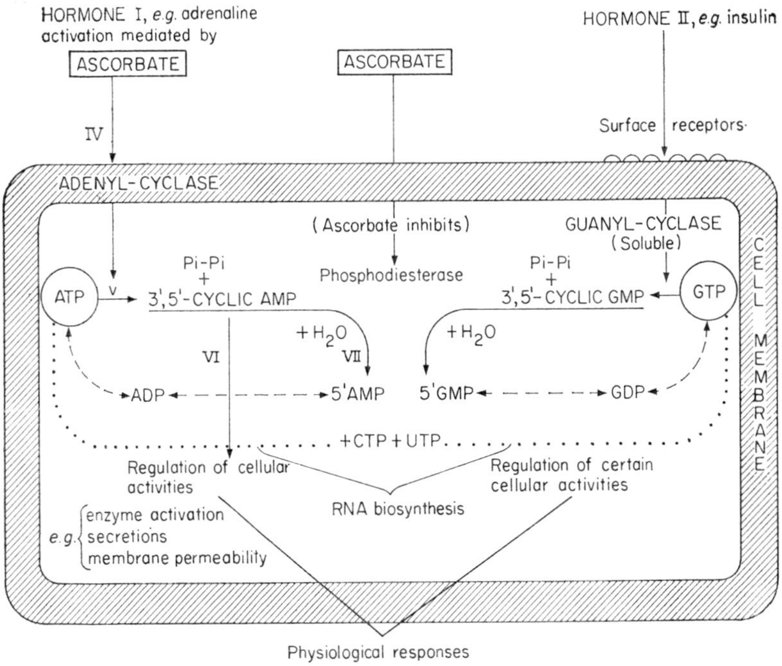

FIG. 5b. Schematic representation of the respective ascorbate enhanced formation and increased activities of cyclic AMP and cyclic GMP and associated physiological effects. (The scheme includes the pathways of formation of the corresponding triphosphates from their respective mono- and diphosphates and relevant involvement of RNA biosynthesis.)

(*i*) *Vitamin C and cyclic AMP*†

(*i-i*) *Adrenaline synthesis and subsequent formation of cyclic AMP*

(1) *Protected formation of adrenaline.* Vitamin C assists the production of adrenaline from dopamine. The activity can be represented schematically as in Fig. 5a.

Stage I. Ascorbate or catechol is required as a co-factor by dopamine β-hydroxylase in the formation of noradrenaline from dopamine.[58]

Stage II. Adrenaline is formed from noradrenaline by methylation. A limited concentration of noradrenaline–methyl–transferase could restrict synthesis of adrenaline, but ascorbate may also be utilised in this stage since Kirshner and Goodall[59] utilised reduced glutathione in their preparations; and reducing thiol activity can be replaced by ascorbate (*see* for example Ref. 45).

Stage III. Protection of adrenaline in the reduced 3′:4′-dihydroxy state. Under physiological conditions adrenaline can be oxidised rapidly to the corresponding dioxy compound which is highly unstable. However, ascorbate rapidly reduces the oxidised compound to the reduced state. Earlier on, in Fig. 3, the association of ascorbate with adrenaline—via the 3′:4′-hydroxy groups—was considered feasible; such association would protect adrenaline from oxidation.

(2) Adrenaline potentiates the adenyl cyclase in the membrane (Stage IV in the overall scheme) which in turn activates the formation of cyclic AMP from ATP (Stage V in scheme).

The last two stages have been elucidated in the literature; for extensive references and tabulation *see* Ref. 37. It is proposed here that the adrenaline–ascorbate complex activates adenyl cyclase either on release of the adrenaline or possibly even directly. In the latter case the 3′ and 4′ hydroxy groups of the adrenaline would be masked.

(*i–ii*) *Inhibition of the hydrolytic activity of PDE on cyclic AMP.* PDE hydrolyses cyclic AMP (Stage VII). It is relevant that cyclic AMP and ascorbate each possess a ring-structure with a negatively charged oxygen. This parallelism and the production of an acid group, when each respective ring is hydrolysed, are indicative of the possibility that ascorbate would compete with cyclic AMP for the enzyme's active sites when the two are present in a total quantity

† A preliminary communication concerning *part* of the scheme was given at the 545th Meeting of the Biochemical Society, 18–19th December, 1973.[56]

sufficient to over-saturate the PDE thereby causing decreased hydrolysis of cyclic AMP. Indeed, Moffat *et al.*[60] have shown that L-ascorbate is effective in concentrations as low as 10^{-4}M in inhibiting the breakdown of cyclic AMP. We have confirmed these observations at $0{\cdot}8 \times 10^{-4}$M ascorbate.[1]

(*i–iii*) *Activity of cyclic AMP in relation to PDE and ascorbate.* Cyclic AMP potentiates hormonal actions, synthesis of a number of enzymes and of several physiological activities (for original references *see* Table 5.1 in Robinson *et al.*[38]) in proportion to its concentration.

As outlined, and shown in Fig. 5b, the concentration level of cyclic AMP depends on its formation and destruction, in

(1) activation of formation from ATP following adenyl-cyclase activity which in turn is potentiated by adrenaline, and
(2) reduction of the activity of PDE which catalyses hydrolysis to 5′AMP.

As the anionic segments of cyclic AMP and of ascorbate are very similar it is likely that their respective association constants with PDE would be of the same order.

Cellular concentration of cyclic AMP can vary from *ca.* 10^{-6} to 10^{-4}M. If the cellular ascorbate ion concentration can reach these or higher levels, it is likely to compete effectively with and displace cyclic AMP from the active sites in PDE, thus tending to inhibit hydrolysis of cyclic AMP and preventing reduction in its concentration.

Now, intake of mega quantities of vitamin C has been shown by several workers to raise the levels of ascorbate concentration in the blood from *ca.* 0·8 mg to 5 mg per 100 ml of blood.† An average value of 2 mg of ascorbate per 100 ml of serum would be equivalent to over 10^{-4}M ascorbate. However, the concentration of ascorbate in the serum is representative only of the *lowest* values, because ascorbate transfer mechanisms exist which pump it into a number of tissues. It has been established by numerous investigators that many tissues have much more ascorbate. For several references *see*

† Different workers give different values, *e.g.* Masek and Hrubá,[61] Kubler and Gehler,[62] Spero,[63] and Coulehan *et al.*,[64] but in general there is agreement that increase in vitamin C intake results in higher *blood* content of ascorbate which tails off due to renal threshold. As will be evaluated later, increase in absorption of ascorbate by the tissues continues in spite of the steady state of concentration attained by the blood.

Table 2 in Lewin.[12] These values are higher than the average concentrations of cyclic AMP in several tissues. It is therefore reasonable to postulate that the higher concentrations of ascorbate will result in increased interaction with PDE thereby reducing the latter's association with cyclic AMP and consequent rescue from being hydrolysed to 5′AMP, thus allowing cyclic AMP concentrations to rise on being formed from ATP. The following experimental findings are in accord with this proposed pattern.

Higher cellular concentrations of cyclic AMP are reflected in increased concentrations in the urine. Thus, Broadus *et al.*[65] found that infusion of adrenaline raises the cyclic AMP levels in both plasma and urine; and Owen and Mofatt (personal communication) noted over 30% increase in the level of cyclic AMP excreted in the urine following several days of daily intake of 2 g of ascorbic acid.

(*ii*) *Cyclic GMP*

(1) *Formation of cyclic GMP, its competitive association with PDE and resultant hydrolysis to 5′-GMP.* When guanyl-cyclase is activated by certain hormones (which do not include adrenaline) GTP is potentiated to form cyclic GMP and pyrophosphate. Ascorbate does not appear to assist hormones which potentiate guanyl-cyclase directly or indirectly—via surface receptors—as is the case with insulin. However, the anionic cyclic phosphate group in cyclic GMP, being identical with that in cyclic AMP, is also similar to the anionic section in ascorbate.

Now, it has been established that cyclic GMP and cyclic AMP can be hydrolysed by a single phosphodiesterase;[66] also that cyclic GMP can be a competitive inhibitor of cyclic AMP.[67] In normal physiological conditions cyclic GMP need not compete with cyclic AMP (for association with PDE) and thereby save it from hydrolysis, and vice versa, because the respective concentrations (allowing for their respective association constant values with PDE) are insufficient to saturate PDE. However, as the ascorbate concentration in the tissues is raised on increased absorption, following intake of mega quantities of the vitamin, increased association with the active sites in PDE should take place, thereby leaving correspondingly smaller numbers of sites for association with cyclic AMP and cyclic GMP. It is now that the competition between the two cyclonucleotides—for the few remaining active sites in PDE—should become prominent, thereby favouring their physiological 'mutual antagonism' recently propounded in the literature.[40] Perhaps the major significance of this aspect lies in the indications that particular balanced concentration ratios of [(cyclic AMP)/(cyclic GMP)] may control some types of malignant tissue development more effectively than cyclic AMP alone[44] without adversely affecting physiologically essential activities.

In this connection it is relevant that the two cyclic nucleotides can compete with one another for particular binding protein. Indeed, Crozier *et al.*[68] have concluded using cyclic AMP binding proteins from bovine adrenal glands that cyclic GMP can displace cyclic AMP from its association with the binding proteins. It is to be expected that ascorbate will tend to displace both cyclonucleotides from their association with these binding proteins.

GENERAL DISCUSSION

It is worthwhile stressing that the overall framework (of the potential of high intake of vitamin C) which comprises both direct and indirect effects, although in accord with current knowledge, may have to be modified in the light of more information. Its usefulness lies in its ability to offer a pattern to many experimental data which cannot otherwise be satisfactorily accommodated; further it enables a more realistic phasing of future research projects.

Accepting the concept that the main actions of intake of mega quantities of vitamin C are physiologically beneficial, a number of persistent questions come to the fore. These questions and related explanations will be briefly tackled as follows.†

(I) Is There any Maximum Level of Intake of Mega Quantities of Vitamin C where its Transfer from the Plasma to the Tissues is no Longer Quantitatively Significant?

It is often argued by opponents of intake of mega quantities of vitamin C that plasma saturation levels are readily attained with a daily dose of about 100 mg and consequently the remainder of the intake of the vitamin is excreted via the urine. This argument assumes that when the plasma is 'saturated' with vitamin C, the other tissues being in equilibrium with it are bound to be also saturated. This argument is naive, since many stereochemically and energetically

† A more extensive examination is undertaken elsewhere.[1]

admissible mechanisms can be formulated in which a *steady* level in the plasma—despite increasing intake of vitamin C—represents a *steady state* in which the increasing ingestion of vitamin C results in simultaneous increasing transfer to the tissues and increasing excretion via the urine. This can be represented diagrammatically as

Increasing intake of vitamin C (I) → Increasing uptake by tissues (T)
→ Steady plasma concentration (S)
→ Increasing excretion via urine (U)

where I = U + T + S.

Increasing uptake by the tissues with increasing ingestion of the vitamin—associated with a *steady state in the plasma*—is particularly likely to take place because of the existence of special anti-concentration gradient mechanisms which operate in a number of tissues where the concentration of the vitamin greatly exceeds that in the plasma (*e.g.* leucocytes and adrenals). Experimental evidence does exist that increasing intake of the vitamin—taken in equal multi-daily doses—results in increasing uptake by the tissues.

Kubler and Gehler[54] have shown that increased ingestion of vitamin C over the range of 1·5–12 g daily results in increasing absorption of the vitamin although the percentage ratio of (absorbed/ingested) material decreases. This can be appreciated from Fig. 6 which is based on their work and on Mayersohn's subsequent re-analysis.[69]

It follows that the significance of the reports that the level of vitamin C in plasma tends to reach a maximum level at *ca.* 1·8 mg per 100 ml (*e.g.* Friedman *et al.*[70]) should not be extended to the implication that once this level is attained, tissue saturation is achieved and that this would happen on intake of *ca.* 100 mg of vitamin C by most individuals (*e.g.* Goldsmith[5]).

(II) Are there any 'Side' Effects which have not as yet been Concisely Appreciated? Can such 'Side' Effects Result in Adverse Effects on Human Physiology, Under Certain Sets of Conditions?

Currently three aspects appear prominent.

(i) Oxalate formation

One recognised effect with wide differentiation in respect of biochemical individuality is that of the formation of oxalates (*see* Fig. 1). This merits only a note in passing because of the extensive

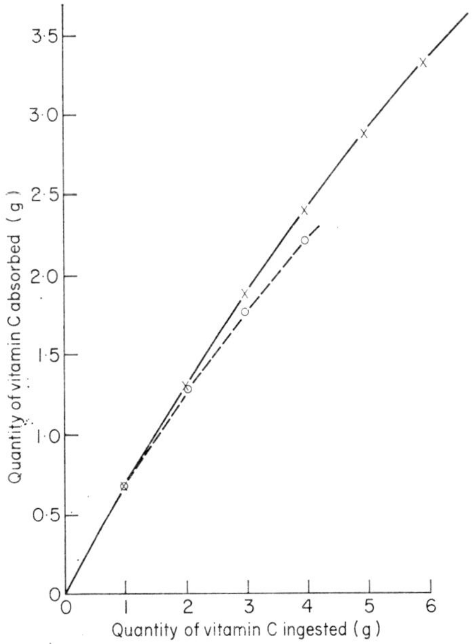

FIG. 6. Variation of the amount of vitamin C absorbed with the quantity ingested. (Plotted from data by Kübler and Gehler[54] and recalculations by Mayersohn.[69]) ×—×, total of 6 g in six equally distributed doses; ○— —○, total of 4 g in four equally distributed doses.

coverage already accorded to it in the literature.[71-74] The formation of oxalic acid increases in a number of individuals with rise in vitamin C intake. It has emerged that this is not general, since in many cases the effect could not be demonstrated;[74] further, a total intake of less than 4 g daily of vitamin C does not raise the oxalate to non-acceptable levels even in individuals who display increased oxalate formation.

Digestive upsets have been recorded in a number of cases when as much as a *single* dose of 3 g of the vitamin has been taken as sodium ascorbate or in the form of tablets (not powder dissolved in grapefruit or orange juice). It is relevant to point out that (1) binding agents used in some vitamin C *tablets* can upset individuals with particular susceptibilities, and (2) intake of large single doses of sodium ascorbate can result in effects which may be attributed to gastric intestinal upsets. This could be due to an antacid reaction. 3 g of vitamin C in the form of sodium ascorbate are equivalent to about 1·5 g of sodium bicarbonate. Certainly some individuals taking 1·5 g of sodium bicarbonate will experience what to them appears to be a stomach upset.

(*ii*) *Lowering of blood sugar*

Quite a number of findings have been recorded that the level of blood sugar is reduced—albeit temporarily—following more than 0·5 g intake of ascorbic acid (*e.g.* Secher[50]; Sylvest[51]). While useful in therapy of hyperglycaemic conditions, it could adversely affect individuals with a tendency to hypoglycaemia. This field deserves a thorough examination.

(*iii*) *Loss of nutritionally required cations*

It has been pointed out here and elsewhere[12] that the ascorbate ion is accompanied by counterions on excretion via the urine. The possibility that the levels of certain nutritionally required cations may be reduced to physiologically unacceptable levels arises. This field is currently undergoing experimental investigation.[76]

(III) Are there any Indications of Subsequent Biosynthetic or Genetic Developments the Effects of which have so far not been Examined in the Literature?

To be in a position to answer this searching question unambiguously it is desirable to evaluate various mechanisms in which the ascorbic system may be involved when it reaches particular concentrations in the cytoplasm and in the nucleus. It is also desirable to evaluate the mechanisms by which ascorbate concentration in certain tissues is raised well above the level in the surrounding inter-cellular fluid. The latter anti-gradient increase in concentration involves an increase in free energy (or decrease in entropy) which must be financed by linkage to other reactions which provide overriding energy outputs. Such required energy outputs are computable, but the mechanisms may vary from tissue to tissue; and exploration of this field is as yet incomplete. To indicate, it has been ascertained that ascorbate penetrates certain membranes such as the erythrocyte membrane much more slowly than DHA (Hughes and Maton[75]). However, these findings have not taken into consideration the potential of cyclic AMP. The latter potentiates increasing membrane permeability and is therefore likely to enhance ascorbate penetration of cells or nuclei whose membranes are normally impermeable to vitamin C. Now, it has been noted that ascorbic acid inactivates PDE and therefore results in increased concentrations of cyclic AMP.[60] Hence, sufficient rise in ascorbate concentration could enhance membrane permeability in particular tissues thereby gaining entry to loci from which it has been previously almost debarred at lower concentrations, where its effect on PDE was insufficient to raise the level of cyclic AMP to that required to potentiate sufficiently high permeability levels. It is therefore necessary to ascertain the association constant values of ascorbate with PDE (optimal pH values *ca.* 7·5–8) and relevant rises in cyclic AMP levels—a project currently being carried out (Lewin[76]).

It is worthwhile emphasising that both ascorbate and cyclic AMP are effective in lowering interfacial tension. It has been computed theoretically that ring-structures tend to reduce the entropy of associated/involved water molecules—as compared to bulk-water—from three-dimensions to lower dimensions, and as a result should reduce the interfacial tension of aqueous solutions.[26] In accord with this evaluation are the experimental results that ascorbic acid/ascorbate reduce the interfacial tension of aqueous solutions by several dynes (Lewin,[1] Künzel[28]) and that cyclic AMP also reduces effectively the interfacial tension of its aqueous solutions (Lewin and Marshall[77]).

The ability to reduce interfacial tension is of fundamental importance in adversely affecting conformations such as those of proteins (which rely on hydrophobic contributions for helical stability) and in membrane permeability (in which hydrophobic associations tend to ensure decreased permeability). This is so because the contribution of hydrophobic adherence increases with rise in interfacial tension according to the equation[26,36,37]

$$L \cong (3 \times 10^{-3} \gamma A) \text{ kcal/mol/mol}$$

where γ = interfacial tension and A is the hydrophobic area of contact. This gives the decrease in free energy (L) arising out of the loss of water interface.

On this basis, increasing concentrations of cyclic AMP—enforced by raised ascorbate concentrations or by other means—should enhance increased permeability to cyclic AMP itself as well as to ascorbate. Higher concentrations of cyclic AMP within the nucleus could give rise to two different but nevertheless complementary effects, namely:

(*a*) *ionic competition* for the positively charged groups of the (nuclear) proteins thereby causing interference with the ionic linkages between the positively charged groups of the proteins and the negatively charged groups of DNA, as well as inter- and intra-protein ionic linkages, and

(*b*) *lowering of the interfacial tension* thereby causing interference with the hydrophobic adherences of protein–DNA and protein–protein.

Increasing concentrations of cyclic AMP within the nucleus might therefore affect chromatin structure by adversely affecting existing ionic linkages and hydrophobic adherences which contribute to the masking of the genetic template.

Precise computing of the effects on biosynthesis obviously depends on correspondingly formulated experimental projects, and their results. It is relevant, however, to consider that much depends on where the biosynthesis of various RNAs takes place. Some might take place in

(*a*) the DNA *grooves*, the phosphate–deoxy ribose chains of which are lined by ionically-linked proteins, the sidechains of which associate by hydrophobic adherences or ionic linkages, and which thereby exert an umbrella-like masking of the DNA grooves[78,79] (*see* Fig. 7a).

Other RNAs may be synthesised in

(*b*) DNA *segments* in which the phosphate groups are *not* attached to the positively charged groups of proteins, but only attached to inorganic cations such as Na^+ (*see* Fig. 7b).†

It is nevertheless feasible to draw at the present stage approximate outlines of a sequence-pattern of stereochemically and energetically admissible activities of cyclic AMP.

The pK value of the negatively charged phosphate group of cyclic AMP is lower than 1, and is of the same order as the negatively charged phosphate groups of DNA. Hence, if the ionic histone–DNA

† It has been computed (Itshaki[80]; Clark and Felsenfeld[81]) that about 50–60% of the negatively charged phosphate groups of DNA are linked to the positively charged groups of proteins.

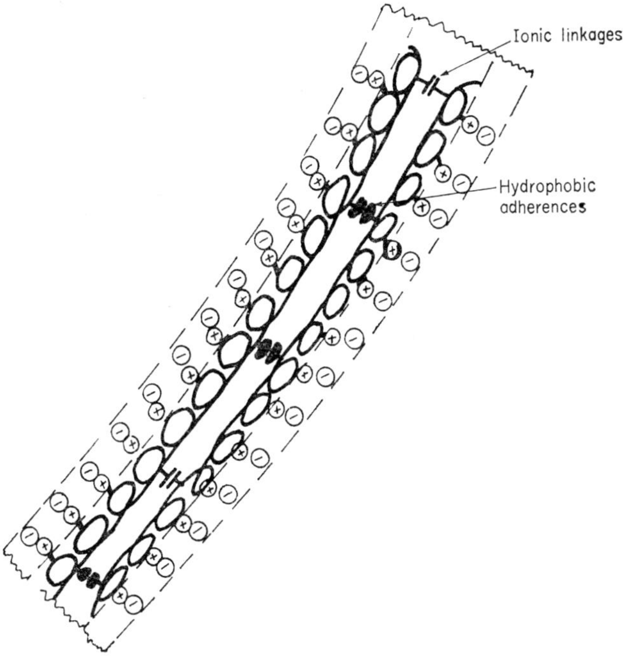

FIG. 7 (a)

linkages were unmasked and available for ionic competition, the cyclic AMP anion would be able to competitively attach itself to the positively charged groups of the proteins (which are ionically linked to the phosphate groups of DNA) thereby tending to detach these proteins from their ionic linkages to DNA. However, the ionic linkages of histones and of acidic proteins to DNA need not be readily penetratable, because they are likely to be masked by extensive inter- and intra-protein associations comprising both ionic linkages and hydrophobic adherences. What is more feasible is interference with the inter- and intra-(nuclear) protein hydrophobic adherences—which should be adversely affected by the interfacial tension-lowering potential of cyclic AMP even at low concentrations such as 10^{-4}M. With increasing concentration of the cyclic nucleotide the competition with inter- and intra-protein ionic linkages should make a further dent in the protein umbrella masking the DNA. It is hoped that current investigations on the effect of ascorbate and of

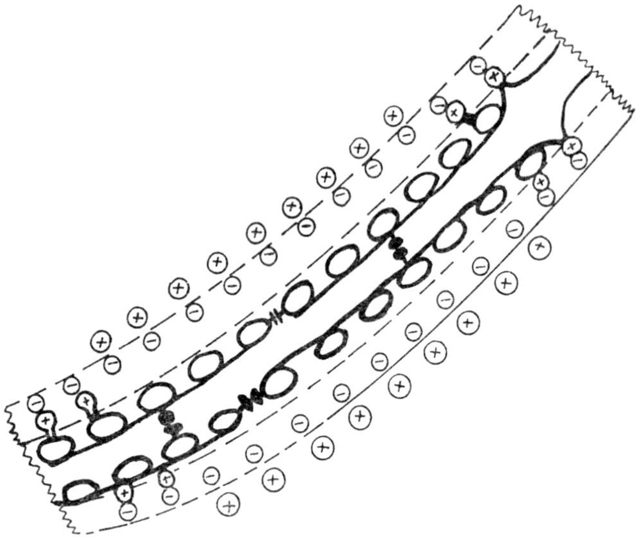

FIG. 7 (b)

FIG. 7. Schematic representation of stereochemically and energetically admissible protein–DNA associations in chromatin in which inter-protein associations can take place via ionic links (*e.g.* carboxylate-δ-ammonium) and hydrophobic adherences. (a) In the DNA grooves the phospho-deoxyribose chains of which are lined by ionically linked proteins to the phosphate groups. (b) In DNA segments in which the phosphate groups are not ionically linked to positively charged groups of proteins, but to inorganic cations such as Na^+.

cyclic AMP on chromatin[77] and on its biosynthetic potential and of ascorbate and cyclic AMP on model systems,[82] such as (poly dA/ poly dT)–histone systems and resultant biosynthesis, will facilitate evaluation of the potential of their direct activities.

At this stage it is worthwhile considering the relative concentrations of the positively charged groups of histones in chromatin and of cellular concentrations of the cyclic AMP anions. While the former is comparatively steady, that of cyclic AMP can vary considerably from one tissue to another and can be raised greatly by adrenaline activity and PDE inactivity. The following evaluation should be of assistance: Turtle and Kipnis[83] determined the concentrations of cyclic AMP in rat pancreatic islets when no exogenous material was added as well as on addition of 5×10^{-5}M adrenaline and of 10^{-3}M theophylline. The corresponding cyclic AMP concentrations varied respectively from *ca.* $3{\cdot}55 \times 10^{-1}$ mol, $4{\cdot}15 \times 10^{-1}$ mol and $55{\cdot}7 \times 10^{-1}$ mol per kg of tissue. Accepting the value of an average concentration of 100–200 g of f_1 histone for a volume of one litre of eukaryotic nucleus (E. W. Johns, personal communication), a molecular weight of *ca.* 20 000 daltons and *ca.* 28 % of positively charged groups for the histone, the concentration of the positively charged groups can be computed to be in the approximate range of $1{\cdot}5 \times 10^{-3}$M to 3×10^{-3}M. In this case the concentration of the cyclic AMP ion greatly exceeds not only the concentration of the positively charged groups of the f_1 histone, but also those of all the other histones present as well as those of the acidic nuclear proteins. However, the concentration of cyclic AMP can be significantly lower in other tissues. Hence, the potential effect of cyclic AMP on RNA synthesis is not only likely to vary from one tissue to another but also within the same tissue depending on factors which raise the nucleotide concentration, such as rise in ascorbate concentration in the cell.

(IV) Are there any Particular Clinical Conditions in which the Link Between Vitamin C and Cyclic AMP, Although Existing, has as yet not been Illuminated?

There are several such conditions, but only one will be briefly discussed here; namely the effect of high intake of vitamin C on mentally depressed conditions. Evidence has been gathered by several investigators (*e.g.* Paul *et al.*;[84,85] Abdulla and Hamadah;[86] Ramsden[87]) that mentally depressed states in humans are associated with below normal cyclic AMP levels, whereas a switch to a maniac state is accompanied by a multifold increase in cyclic AMP concentration levels. Also Milner[88] and Punekar[89] have obtained evidence that the vitamin C levels are lowered in individuals with mentally depressed states. The exact pattern of the relationship is not thoroughly clear, but this may be attributed to variations in cyclic GMP content which would tend to blur the picture. Clarification is likely to be enhanced by a comprehensive investigation of the individual concentrations of cyclic AMP and cyclic GMP in the urine and in the plasma of persons subjected to normal and to high intake of vitamin C and relating these to any associated variations in resistance to body malfunctions. The project is now past the planning stage.[76]

(V) In View of the Possible Co-existence of Several Open-chain Forms Along with the Ring-structured Vitamin C, do the Current Methods of Vitamin C Assay Concisely Evaluate the Amount of Vitamin Present?

Current methods using vitamin C assay can be placed in two categories which depend on corresponding properties of the vitamin, namely:†

(*a*) Reducing ability, and (*b*) reaction with 2:4-dinitrophenylhydrazine to form the corresponding osazone.

There is general agreement that the biologically active form of the vitamin is the γ-lactone structure comprising a double-bond between the α-C and the β-C and—under physiologically active conditions—one ionisable group associated with the double-bond segment.

It follows that if either of these pre-requisites is lacking, vitamin activity is no longer certain. Yet the ring-structure can be hydrolytically opened still leaving the double-bond with the two associated hydroxy groups which are still capable of reduction. The resulting compound shown on the right of Fig. 1‡ cannot be expected to be biologically active and yet it has a reducing potential. The effect of this compound should be allowed for, because of experimental evidence recently obtained[1] that ascorbate can be hydrolysed by PDE giving rise to increased acidity. Work is currently in progress to ascertain the corrections required in concise vitamin assay.

REFERENCES

1. Lewin, S. (1975). *The Molecular Biology and Potential of the Ascorbic System*, Academic Press, London and New York, in the press.
2. Goth, A. and Littmann, I. (1948). *Cancer Research*, **8**, p. 349.
3. Waldo, A. L. and Zipf, R. E. (1955). *Cancer*, **8**, p. 187.
4. Hume, R., Weyers, E., Rowan, T., Reid, D. S. and Hillis, W. S. (1972). *Brit. Heart J.*, **34**, p. 238.
5. Goldsmith, G. A. (1961). *Ann. N.Y. Acad. Sci.*, **92**, p. 230.
6. Pauling, L. (1970). *Vitamin C and the Common Cold*, San Francisco, California.
7. Stone, I. (1972). *The Healing Factor, Vitamin C Against Disease*, Grosset and Dunlap, New York.

† For reviews and extensive references *see* for example Roe.[90,91]

‡ This compound does not spontaneously form the γ-lactone; it requires treatment with 8% HCl at 50°C to form the vitamin.[18]

8. Burch, H. B. (1961). *Ann. N.Y. Acad. Sci.*, **92,** p. 268.
9. Spero, L. (1973). Oral contribution at the Stanford Conference on *Vitamin C and the Common Cold*, 6–7 August, 1973; quoted in Lewin, Ref. 25.
10. Yew, M. S. (1973). *Proc. Nat. Acad. Sci. U.S.*, **70,** p. 969.
11. Coulehan, J. L., Reisinger, K. S., Rogers, K. D. and Bradley, D. W. (1974). *New England J. Med.*, **290,** p. 610.
12. Lewin, S. (1974). *Comp. Biochem. Physiol.*, **47B,** p. 681.
13. Conney, A. H., Bray, G. A., Evans, C. and Burns, J. J. (1961). *Ann. N.Y. Acad. Sci.*, **92,** p. 115.
14. Barnes, M. J. and Kodicek, E. (1972). *Vitamins and Hormones*, Vol. 30, p. 1, Eds. R. S. Harris, P. L. Munson, J. Glover and E. Dkzfalusy, Academic Press, New York.
15. Levin, Y. E. and Kaufman, S. (1961). *J. Biol. Chem.*, **236,** p. 2043.
16. Haworth, W. N. (1933). *Chem. Ind.*, p. 482.
17. Hirst, E. L. (1933). *Chem. Ind.*, p. 221.
18. Haworth, W. N. and Hirst, E. L. (1933). *Chem. Ind.*, p. 645.
19. Herbert, R. W., Hirst, E. L., Percival, E. G. V., Reynolds, R. J. W. and Smith, F. (1933). *J. Chem. Soc.*, p. 1270.
20. Birch, T. W. and Harris, L. J. (1933). *Biochem. J.*, **27,** 595.
21. Karrer, P., Schwarzenbach, K. and Schöpp, G. (1933). *Helv. Chim. Acta*, **16,** p. 302.
22. Borsook, H., Davenport, H. W., Jeffreys, C. E. P. and Warner, R. G. (1937). *J. Biol. Chem.*, **117,** p. 237.
23. Ball, E. G. (1937). *J. Biol. Chem.*, **118,** p. 219.
24. Lewin, S. (1973). *Trans. Biochem. Soc.* (545th Meeting), **1,** p. 154.
25. Lewin, S. (1974). *Chem. Brit.*, **10**(1), p. 25.
26. Lewin, S. (1974). *Displacement of Water and its Control of Biochemical Reactions*, Academic Press, London and New York.
27. Lewin, S. (1972). *Brit. J. Haematol.*, **22**(5), p. 561.
28. Künzel, O. (1941). *Ergebn. Inn. Med. Kinderheilk.*, **60,** p. 565.
29. Pitts, R. F., Lospeosch, W. D., Schiess, W. A. and Ayer, J. L. (1948). *J. Clin. Invest.*, **27,** p. 48.
30. Smith, H. W. (1951). *The Kidney; Structure and Function in Health and Disease*, Oxford University Press, p. 136.
31. Schwab, M. and Kuhns, K. (1959). *Die Storungen des Wasser- und Elektrolytstoffwechsels*, Springer Verlag, Berlin, p. 171.
32. Consolazio, C. F., Johnson, R. E. and Pecora, L. J. (1960). In *Physiological Measurements of Metabolic Functions*, McGraw-Hill, New York, p. 437.
33. Ruskin, A. and Ruskin, B. (1952). *Texas Reports on Biology and Medicine*, **10,** p. 429.
34. Gontze, A. J., Dumitrache, S., Rujinski, A. and Cocora, D. (1963). *Int. Z. Physiol. einschl. Arbeitsphysiol.*, **20,** p. 20.
35. Marchmont-Robinson, S. W. (1941). *J. Lab. Clin. Med.*, **114,** p. 1178.
36. Lewin, S. (1973). *Nature, New Biology*, **231,** p. 80.
37. Lewin, S. (1971). *First European Biophysics Congress*, III, p. 471.

38. Robison, G. A., Butcher, R. W. and Sutherland, E. W. (1971) *Cyclic AMP*, Academic Press, New York and London.
39. Butcher, R. W., Robison, G. A. and Sutherland, E. W. (1970). In CIBA Symposium on *Control Processes in Multicellular Organisms*, J. & A. Churchill, London, p. 64.
40. Goldberg, N. D., Haddox, M. K., Dunham, E., Lopez, C. and Hadden, J. W. (1973). Cold Spring Harbor Meeting on *Proliferation in Animal Cells*, p. 40.
41. Benade, L., Howard, T. and Burk, D. (1969). *Oncology*, **23,** p. 33.
42. Bürk, R. R. (1968). *Nature, Lond.*, **219,** p. 1272.
43. Gericke, D. and Chandra, P. (1969). *Hoppe-Seyler's Z. Physiol. Chem.*, **350,** p. 1469.
44. Ryan, W. L. and Heidrick, M. L. (1968). *Science*, **162,** 1484.
45. Price, C. E. (1966). *Nature, Lond.*, **212,** p. 1481.
46. Emmer, M., de Crombrugghe, B., Pastan, I. and Perlman, R. (1970). *Proc. Nat. Acad. Sci. U.S.*, **66,** p. 480.
47. Earp, H. S., Watson, B. S. and Ney, R. L. (1969). *Clin. Res.*, **17,** p. 22.
48. Friedman, R. M. and Pastan, I. (1969). *Biochem. biophys. Res. Commun.*, **37,** p. 735.
49. Illiano, G., Tell, G. P. E., Siegel, M. I. and Cuatrecasas, P. (1973). *Proc. Nat. Acad. Sci. U.S.*, **70,** p. 2443.
50. Secher, K. (1942). *Acta Med. Scand.*, **60,** p. 255.
51. Sylvest, O. (1942). *Acta Med. Scand.*, **60,** p. 183.
52. Pfleger, R. and Scholl, F. (1937). *Wiener Arch. Innere Mediz.*, **31,** p. 219.
53. Bartelheimer, H. (1939). *Mediz. Welt*, **13,** p. 117.
54. Kubler, W. and Gehler, J. (1970). *Intern. Z. Vitaminforsch.*, **40,** p. 442.
55. Dice, J. F. and Daniel, C. W. (1973). *IRCS Med. Sci.* (73/3). 10-19-1.
56. Lewin, S. (1973). 545th *Meeting Biochem. Soc.* p. 22; (1974). *Trans. Biochem. Soc.*, **2**(3), p. 400.
57. Ball, E. G., Chen, T. T. and Clark, W. M. (1933). *J. Biol. Chem.*, **102,** p. 691.
58. Levin, Y. E. and Kaufman, S. (1961). *J. Biol. Chem.*, **236,** p. 2043.
59. Kirshner, N. and Goodall, McC. (1957). *Biochim. Biophys. Acta*, **24,** p. 658.
60. Moffat, A. C., Patterson, D. A., Curry, A. S. and Owen, P. (1972). *Eur. J. Toxicol.*, **5**(3), p. 160.
61. Masek, F. and Hruba, J. (1964). *Intern. Z. Vitaminforschung.*, **34,** p. 39.
62. Kubler, W. and Gehler, J. (1970). *Intern. Z. Vitaminforschung.*, **40,** p. 442.
63. Spero, L. (1973). In a paper presented at the Stanford Conference on *Vitamin C and the Common Cold*, 6–7 August, 1973; quoted in Lewin, Ref. 25.
64. Coulehan, J. L., Reisinger, K. S., Rogers, K. D. and Bradley, D. W. (1974). *New England J. Med.*, **290,** p. 6.
65. Broadus, A. E., Kaminsky, N. I., Hardman, J. G. Sutherland, E. W. and Liddle, G. W. (1970). *J. Clin. Invest.*, **49,** p. 2222.

66. Beavo, J. A., Hardman, J. G. and Sutherland, E. W. (1970). *J. Biol. Chem.*, **245,** p. 5649.
67. Goren, E., Erlichman, J., Rosen, M. and Rosen, S. M. (1970). *Fed. Proc.*, **29,** Part I, p. 602, Abstract No. 1995.
68. Crozier, D. H., Dickinson, J. R. and Swoboda, B. E. P. (1973). 545th *Biochem. Soc. Meeting*, p. 37; (1974). *Trans. Biochem. Soc.*, **2**(3), p. 415.
69. Mayersohn, M. (1972). *Eur. J. Pharmacol.*, **19,** p. 140.
70. Friedman, G. J., Sherry, S. and Ralli, E. P. (1940). *J. Clin. Invest.*, **19,** p. 685.
71. Lamden, M. M. (1971). *New England J. Med.*, **284,** 336.
72. Briggs, M. H., Garcia-Webb, P. and Davies, P. (1973). *Lancet*, **ii,** p. 201.
73. Smith, L. H., Fromm, H. and Hoffmann, A. F. (1972). *New England J. Med.*, **286,** p. 1371.
74. Poser, E. (1972). *New England J. Med.*, **287,** p. 412.
75. Hughes, R. E. and Maton, S. C. (1968). *Brit. J. Haematol.*, **14,** p. 247.
76. Lewin, S. Part of the current research program. Unpublished.
77. Lewin, S. and Marshall, P., manuscript in preparation.
78. Lewin, S. (1970). *J. theoret. Biol.*, **29,** p. 1.
79. Lewin, S. *Biopolymer Structure and Function Potential,* manuscript in preparation.
80. Itshaki, R. F. (1970). *Biochem. Biophys. Res. Commun.* **41,** p. 25.
81. Clark, R. J. and Felsenfeld, G. (1971). *Nature* (*New Biology*), **229,** p. 101.
82. Lewin, S. and Stubbs, G. current research project. Unpublished.
83. Turtle, J. R. and Kipnis, D. M. (1967). *Biochem. Biophys. Res. Commun.*, **28**(5), p. 797.
84. Paul, M. I., Ditzion, B. R. and Janowsky, D. S. (1970). *Lancet*, **i,** p. 88.
85. Paul, M. I., Cramer, H. and Bunney, W. E. (1971). *Science*, **171,** p. 300.
86. Abdulla, Y. H. and Hamadah, K. (1970). *Lancet*, **i,** p. 378.
87. Ramsden, E. N. (1970). *Lancet*, **ii,** p. 108.
88. Milner, G. (1963). *Brit. J. Psychiat.*, **109,** p. 294.
89. Punekar, B. D. (1961). *Ind. J. Med. Res.*, **49,** p. 828.
90. Roe, J. H. (1954). *Methods of Biochemical Analysis*, Vol. 1. Ed. D. Glick, Interscience, New York, p. 115.
91. Roe, J. H. (1961). *Ann. N.Y. Acad. Sci.*, **92,** p. 277.

DISCUSSION

Rivers: Dr. Lewin's lecture confirms my suspicion that molecular biologists are thick on theory but thin on evidence. Does he have any evidence that cyclic AMP or GMP levels are increased in people taking megavitamin quantities of vitamin C?
Lewin: Studies have shown (I can give you the reference afterwards) that the higher the cyclic AMP level in the blood the more cyclic AMP is excreted in the urine. Moffatt and Owen have shown that if you take about 2 g of ascorbic acid the result is increased concentration of cyclic AMP in

the urine. The total daily output is increased. I have checked these data and found it is correct, but they have done most of the work.

Coultate: Are the kinetic parameters for the interaction of ascorbic acid with adrenalin metabolism compatible with a physiological significance for ascorbic acid in this context?

Lewin: Yes. One point I should have mentioned is that if you calculate 1·4 mg of ascorbic acid per 100 mg of serum, you find it is of the order of 10^{-4} M in the serum. Now, normally, cyclic AMP and cyclic GMP find enough space in the phosphodiesterase without affecting one another—in other words, they can go on being hydrolysed without competing for space. If, however, you add 10^{-4} M ascorbate, it doesn't saturate, but it takes up anything from about half to three-quarters of the phosphodiesterase, and then the cyclic AMP and cyclic GMP start competing and this is why their level goes up. I think this is the fundamental point about the ascorbate, because if the ascorbate level starts decreasing, that competition doesn't proceed, and as a result cyclic AMP begins to be hydrolysed much more quickly, and cyclic GMP as well, and their levels start decreasing, and your patient begins to feel the effect—which the clinicians would describe as ascorbic acid deficiency correctly, but without taking any account of what happens at the molecular level.

Coultate: Is anyone aware, in view of the reported disorders of gut function which may follow large doses of vitamin C, of any data on the microbial metabolism of vitamin C?

Lewin: Can I answer you (not so far as the microbial part is concerned) by giving you a very simple calculation? Many a patient who has taken what he calls 3 g of ascorbic acid has in fact taken 3 g of ascorbic acid in the form of sodium ascorbate. Now if you do a simple calculation you will find that 3 g of ascorbic acid in the form of sodium ascorbate contains about the equivalent of 1·5 g of sodium bicarbonate.

Index